Experimental Mechanics of Solids

28th SYMPOSIUM ON EXPERIMENTAL MECHANICS OF SOLIDS
in memory of prof. Jacek Stupnicki
October 17-20, 2018
Jachranka near Warsaw

Editors

Paweł Pyrzanowski, D.Sc., Ph.D., WUT Professor
Mateusz Papis, M.Sc

Institute of Aeronautics and Applied Mechanics
Faculty of Power and Aeronautical Engineering
Warsaw University of Technology
24 Nowowiejska str., 00-665 Warsaw, Poland

Peer review statement

All papers published in this volume of "Materials Research Proceedings" have been peer reviewed. The process of peer review was initiated and overseen by the above proceedings editors. All reviews were conducted by expert referees in accordance to Materials Research Forum LLC high standards.

Published under License by **Materials Research Forum LLC**
Millersville, PA 17551, USA

Published as part of the proceedings series
Materials Research Proceedings
Volume 12 (2019)

ISSN 2474-3941 (Print)
ISSN 2474-395X (Online)

ISBN 978-1-64490-020-8 (Print)
ISBN 978-1-64490-021-5 (eBook)

This book contains information obtained from authentic and highly regarded sources. Reasonable efforts have been made to publish reliable data and information, but the author and publisher cannot assume responsibility for the validity of all materials or the consequences of their use. The authors and publishers have attempted to trace the copyright holders of all material reproduced in this publication and apologize to copyright holders if permission to publish in this form has not been obtained. If any copyright material has not been acknowledged please write and let us know so we may rectify in any future reprint.

Distributed worldwide by

Materials Research Forum LLC
105 Springdale Lane
Millersville, PA 17551
USA
http://www.mrforum.com

Manufactured in the United State of America
10 9 8 7 6 5 4 3 2 1

Table of Contents

Preface

The 28th Polish Symposium on Experimental Mechanics of Solids was held October 17-20, 2018 in Jachranka, near Warsaw. It was organised on behalf of the Institute of Aeronautics and Applied Mechanics, Warsaw University of Technology; Committee on Mechanics of the Polish Academy of Sciences and Polish Association for Experimental Mechanics. The conference is organised every 2 years, since 2006 it is in memory of prof. Jacek Stupnicki - one of the most known polish scientists in the field of experimental mechanics.

The main purpose of the Symposium is to enable researchers to present their latest experimental achievements in mechanics of solids, machine design, mechanical engineering, biomechanics etc.

In 2018 the conference was attend by more than 60 participants. Best papers were selected by the Scientific Committee for full-length publication.

Organized on behalf of:

Warsaw University of Technology

Institute of Aeronautics and Applied Mechanics,
Warsaw University of Technology

PAN

POLISH ACADEMY OF SCIENCES

Committee on Mechanics of the Polish Academy of Sciences

PSME

Polish Association for Experimental Mechanics

Committees

HONORARY COMMITTEE
Marek Bijak-Żochowski
Lech Dietrich
Małgorzata Kujawińska
Józef Szala

SCIENTIFIC COMMITTEE
Paweł Pyrzanowski - Chairman
Romuald Będziński
Dariusz Boroński
Aniela Glinicka
Jerzy Kaleta
Maria Kotełko
Zbigniew Kowalewski
Grzegorz Milewski
Tadeusz Niezgoda
Leszek Sałbut

ORGANISING COMMITTEE
prof. Paweł Pyrzanowski - Chairman
Irena Mruk, Ph.D. – Honorary Secretary
Witold Rządkowski, Ph.D. – Secretary
Michał Kowalik, Pd.D. – Secretary
Dawid Maleszyk, M.Sc. – Secretary
Mateusz Papis, M.Sc. – Secretary

Sponsors

Symposium co-financed by the Polish Academy of Sciences
as part of financing of dissemination activities for science

Experimental Mechanics of Solids
Materials Research Proceedings **12** (2019) 1-8

Materials Research Forum LLC
https://doi.org/10.21741/9781644900215-1

Mechanical Properties Investigation of Composite Sandwich Panel and Validation of FEM Analysis

Szymon Jakubiak[1,a], Filip Ćwikła[2,b], Witold Rządkowski[3,c *]

[1-3] Faculty of Power and Aeronautical Engineering, Warsaw University of Technology, Nowowiejska 24, 00-665 Warsaw, Poland

[a]szymonjakubiak@wutracing.pl, [b]filipcwikla@wutracing.pl, [c]witold.rzadkowski@pw.edu.pl

Keywords: Composites, Carbon Fibers, Epoxy Resin, Tensile Test, Aluminum Honeycomb Sandwich

Abstract. Composite sandwich structures allows for significant mass reduction compared with traditional steel frames used in Formula Student bolides. However, in order to reach a full potential of composite monocoque chassis and ensure safety of the driver, it is required to perform physical tests of sandwich structure and its elements, supported with FEM analysis. Mechanical properties of sandwich panel are affected by a variety of factors, including but not limited to, manufacturing technique and curing conditions. Therefore, series of experiments were performed to determine in-plane tensile, compressive and shear properties of skin material fabricated in conditions related to those projected for monocoque. The acquired data was compared with expected results and used in further FEM analysis. Apart from uniaxial tensile tests, the whole sandwich structure, was tested in 3-point bending and "punch through" strength. A FEM model of each test was created in order to validate a data from more complex simulations.

Introduction

Mechanical properties of CFRP composites depend heavily on performed manufacturing process. Therefore, a series of tests is essential in order to achieve reasonable design, without oversizing the structure.

Main parameters required in design of composite monocoque for Formula Student competition were Young's modulus of laminates, energy absorbed by composite sandwich panel and perimeter shear strength of skin material.

Similar studies of laminates [1,2] and sandwich panels [3-6] were conducted in the past.

Bending and shear tests were performed in accordance with Formula SAE Rules[7] in regards to side impact structure requirements, i.e. punch through strength and energy absorption in bending.

Methods

All tests were performed on Instron 8516 Testing System.

Young's modulus of laminates was derived from uniaxial tensile test. All specimens were prepared in accordance with ASTM 3039[8]. E-glass tabs were laminated to each carbon fibre laminate.

First group of test samples was made of Tenax UTS50 12K F22 S carbon fibre tow infused during wet lay-up with Havel epoxy resin LH 385.

Wet lay-up is a method used for manufacturing wide range of components for Formula Student bolides e.g. front and rear wings, diffuser and sidepods.

Experimental Mechanics of Solids

Materials Research Forum LLC

Materials Research Proceedings **12** (2019) 1-8

https://doi.org/10.21741/9781644900215-1

Table 1. Configuration of samples from wet lay-up

lay-up	number of specimens
[0]	8
[0$_2$]	12

Second group of specimens was manufactured in prepreg technology from KORDCARBON-CPREG-200-P-3K-EP1-42, cured in 120°C in a vacuum bag under atmospheric pressure.

Prepreg technology is used for manufacturing composite monocoque of the bolide. This method allows for cutting time of manufacturing and increasing consistency of properties.

Table 2. Configuration of prepreg samples

lay-up	number of specimens
[0]	1
[0$_3$]	3
[45$_3$]	1
[0/45/0]	1

A load-displacement curve was obtained directly from testing machine. Resulting strain was compared with strain from Digital Image Correlation, necessary calibration was introduced.

All specimens were covered in stochastic pattern and then recorded in UHD 30fps throughout the whole tensile test. With the use of extensometers in GOM Correlate software, longitudinal strain was calculated for every frame of recorded video.

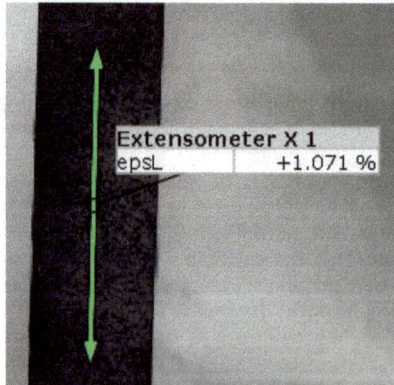

Fig. 1. Extensometer created from 2 points based on stochastic pattern

Rectangular 275 mm x 500 mm composite sandwich panels were tested in 3-point bending. Panels were supported by a span distance of 400mm. Energy absorbed in total displacement of 19 mm was compared with energy absorbed by two 25 mm x 1.75 mm round steel tubes bended on the same test rig, to prove structural equivalency of composite side impact structures in bolide with monocoque. Steel force applicator with 50mm radius overhang test sample at both ends in order to prevent edge loading.

Experimental Mechanics of Solids   Materials Research Forum LLC
Materials Research Proceedings 12 (2019) 1-8   https://doi.org/10.21741/9781644900215-1

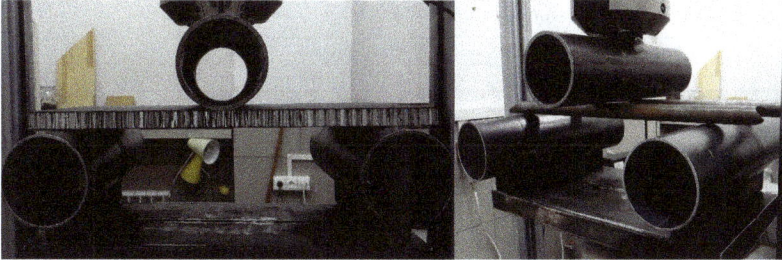

Fig. 2. Composite (left) and steel (right) samples placed on the test rig

Sandwich panels cores (20 mm thick) were made of 3003 aluminium alloy. Cell size of the honeycomb was 6.4mm with nominal foil gauge of 0.075mm. Material used for skin was Kordcarbon prepreg. For interface between core and skin materials, XPREG XA120 Prepreg Adhesive Film was used. All test samples were symmetric i.e. fibre orientation in skins was symmetric about panel midplane.

Table 3. Configuration of samples for 3-point bending

sample id	skin lay-up	adhesive film
1b	[0/45]	no
2b	[0/45/0]	yes
3b	[0/45/0/45]	yes

Perimeter shear strength was derived from first highest force peak in punch through test, in which a steel Ø25 mm punch was pushed against a sandwich panel supported by steel plate with cylindrical Ø32 mm hole for the punch.

Fig. 3. Setup for punch-through test

Square 100 mm x 100 mm panel samples were manufactured under the same conditions and from same materials as specimens for 3-point bending. The aim of this test is to provide data for inserts design and to ensure safety of the driver in case of collision with a random debris on a track.

Table 4. Configuration of samples for punch-through test

sample id	skin lay-up	adhesive film
2p	[45/0/45]	yes
3p	[45/0/45/0/45]	yes
1p	[45/0/45/0/45/0/45]	yes

Results and Discussion

Each sample for tensile test was mounted on tensile test machine and loaded until failure. Average results are presented in Table 5, 6. The data scatter of Young's modulus is acceptable and tested values may be used in further design and FEM analyses.

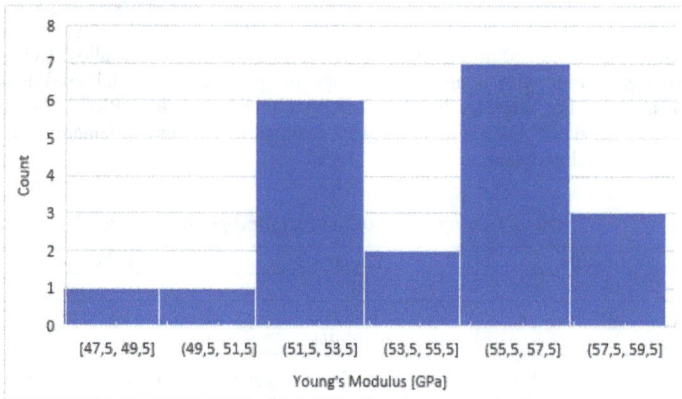

Fig. 4. Distribution of Young's modulus for wet lay-up samples

Table 5. Average results for wet lay-up samples

	Young's Modulus [GPa]	Strain to Failure [%]	Tensile Strength [MPa]
measured	54,7	1,70	740

Table 6. Average results for prepreg samples correlated with data from vendor

	Young's Modulus [GPa]	Strain to Failure [%]	Tensile Strength [MPa]
datasheet	58,6	1,3	778
measured	57,9	1,01	508

Most common failure point of the samples was located near tab ends, Figure 5. Due to stress concentration at this points, measurements of nominal values of ultimate strain and tensile strength for laminates was not possible. However, Young's modulus derived from linear-elastic region meet their expected values.

Experimental Mechanics of Solids   Materials Research Forum LLC
Materials Research Proceedings **12** (2019) 1-8   https://doi.org/10.21741/9781644900215-1

Fig. 5. Failure mode of tensile test specimen

Sandwich panels in 3-point bending shows long liner region until a failure of the skin on compressed side of the sample, Figure 7, 8. Properly estimated stiffness of sandwich panels is crucial in further FEA of monocoque chassis.

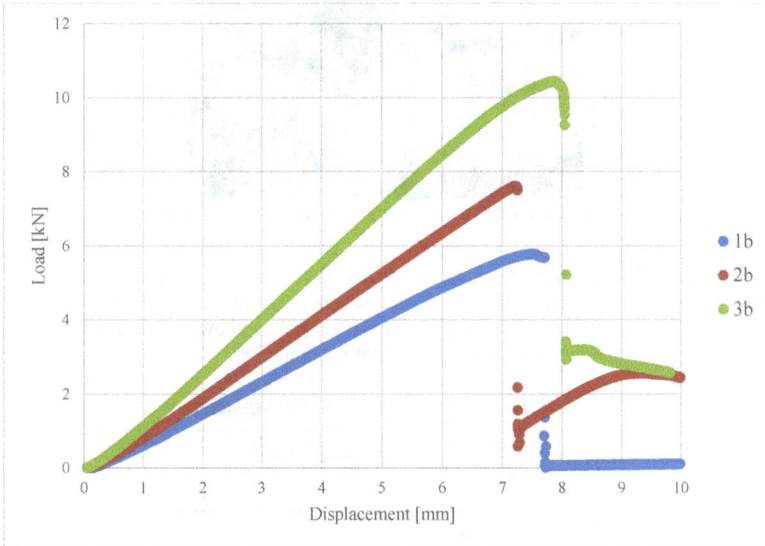

Fig. 6. Load-deflection curve for bended sandwich panels

Fig. 7. Failure of compressed skin in sandwich panel

Fig. 8. Failure of skin in compression and core delamination

Tested panels failure starts from skin compression. Samples with adhesive film keep absorbing energy after first ply failure, while the one without adhesive layer loses its load carrying capacity after skin delamination. This comparison shows an importance of using adhesive film which prevents a skin from premature delamination.

Table 7. Energy absorbed on 19 mm displacement

	1b	2b	3b	tubes
Absorbed energy [J]	23	43	59	96

Based on results from tensile test a FEM model of 3-point bending was created. The shell elements allows for further scaling of the simulation up to the analysis of the whole monocoque structure leaving computing power requirements on a reasonable level. The boundary conditions (Figure 9) come down to fixed support tube (B), 0 displacement on one edge of a panel (A) and respective displacement on force applicator resulting in the highest load measured during the bending test (C). Contact between a panel and force applicator was set to frictionless while for support tube there was fictional contact with a friction coefficient of 0,2.

Experimental Mechanics of Solids

Materials Research Forum LLC

Materials Research Proceedings **12** (2019) 1-8

https://doi.org/10.21741/9781644900215-1

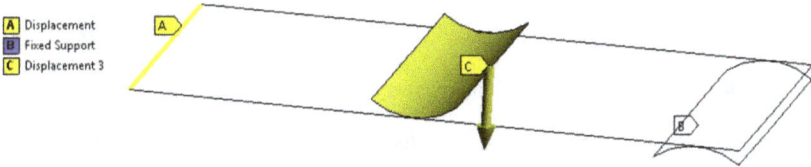

Fig. 9. Boundary conditions for a panel in 3-point bending

With global element size of 10mm the accuracy of discretization was sufficient. In comparison to 2,5mm element size there is 5% difference in results. Critical stress values for FEA were calibrated on b3 case so that it indicates first ply failure during the highest load from physical test. This approach results in inverse reserve factors for b1 and b2 cases respectively 1,10 and 0,90.

Fig. 10. Tsai-Wu criterion for b3 panel under 10,4kN load

Results from punch-through test show linear relation between thickness of the skin laminate and maximum force recorded during the test. The differences between highest force peaks during punching through the top and the bottom laminate are more significant for samples with higher thickness.

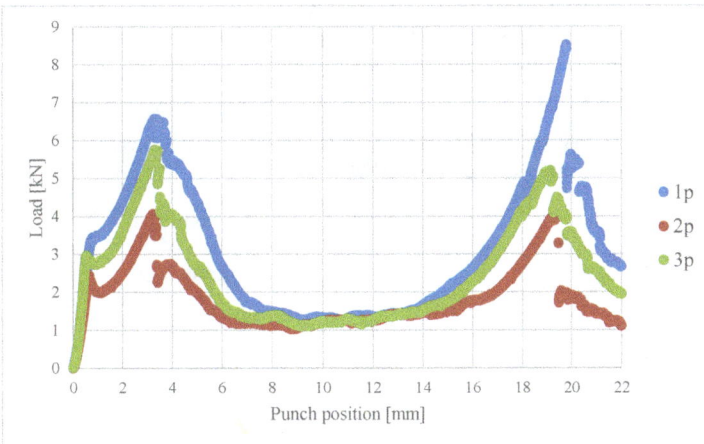

Fig. 11. Load-deflection curve for punch-through test

Conclusion

Tensile test allowed for determination of Young's modulus for different laminates with the accuracy sufficient for design of Formula Student bolide. Further improvement of sandwich panels is required in order to provide higher energy absorption in bending.

Both prepreg and wet laminate present similar values of Young's modulus while UTS50 carbon fibre shows superior strength properties over its prepreg counterpart and should be taken under consideration in design of highly loaded composite parts.

FEA estimation of first ply failure varies by ±10% between different panels in comparison with physically tested values. This deviation may be caused by local stresses introduced by a force applicator or overall scatter in panels properties.

References

[1] J.M.F de Paiva, S. Mayer, M.C. Rezende, Comparison of tensile strength of different carbon fabric reinforced epoxy composites, Mat. Res. vol.9 no.1 São Carlos Jan./Mar. 2006. https://doi.org/10.1590/s1516-14392006000100016

[2] H. Rahmani, S.H.M. Najafi, S.S. Matin, A. Ashori, Mechanical Properties of Carbon Fiber/Epoxy Composites: Effects of Number of Plies, Fiber Contents, and Angle-Ply Layers, Polymer Engineering and Science 54(11) November 2014. https://doi.org/10.1002/pen.23820

[3] G. Sun, X. Huo, D. Chen, Q. Li, Experimental and numerical study on honeycomb sandwich panels under bending and in-panel compression, Materials & Design vol.133, 5 Nov. 2017, pp. 154-168. https://doi.org/10.1016/j.matdes.2017.07.057

[4] D. Ruan, G. Lu, Y.C. Wong, Quasi-static indentation tests on aluminium foam sandwich panels, Composite Structures, 92, (2010), 2039–2046. https://doi.org/10.1016/j.compstruct.2009.11.014

[5] Chen, C., Harte, A.-M., and Fleck, N. A., 2000, The Plastic Collapse of Sandwich Beams With a Metallic Foam Core, Int. J. Mech. Sci., 43, pp. 1483–1506. https://doi.org/10.1016/s0020-7403(00)00069-2

[6] H. Bart-Smith, J. W. Hutchinson, A. G. Evans, , Measurement and Analysis of the Structural Performance of Cellular Metal Sandwich Construction, Int. J. Mech. Sci., 43, pp. 1945–1963. (2011). https://doi.org/10.1016/s0020-7403(00)00070-9

[7] Formula SAE Rules 2019 v2.1, SAE International, pp. 36-37. (2018)

[8] ASTM D3039, Standard Test Method for Tensile Properties of Polymer Matrix Composite Materials, American Society for Testing Materials.

Experimental Mechanics of Solids   Materials Research Forum LLC
Materials Research Proceedings **12** (2019) 9-18   https://doi.org/10.21741/9781644900215-2

24h Creep Test of PE1000 Material at Elevated Temperatures

Bogusz Paweł[1,a*], Grzeszczuk Wojciech[2,b]

[1]Faculty of Mechanical Engineering, Military University of Technology,
00-908 Warsaw, Witolda Urbanowicza Street 2, Poland

[a]pawel.bogusz@wat.edu.pl, [b]wojciech4741@gmail.com

Keywords: Polyethylene, Creep Tests, Mechanical Properties, Experimental Mechanics

Abstract. Plastics, also referred to as polymeric materials, are among basic groups of engineering materials. Creep in plastics, generally, results from the chains straightening, their relative to each other and orientation in the acting force direction. The main factors initiating the occurrence of the rheological phenomena are load and temperature. A wide range of the plastic materials applications includes elements exposed to long-term loads and working at elevated temperatures, which leads to conduct rheological tests. The UHMW (ultra-high molecular weight) type PE1000 polyethylene, in the form of a 4 mm thick plate was tested experimentally. A test program was established using three creep constitutive function variables: temperature, load and time. The maximum duration of the tests was set to 24 hours. Three temperature values, 22°C, 42°C and 90°C, were chosen based on the characteristics of the material. Loading values at creep were selected based on the static tensile tests, which were carried out for each given temperature. The reference point was the material strength of the polyethylene at a given temperature. The samples were stressed with seven multipliers of the ultimate strength, from 0.9 to 0.4, at a given temperature. Creep curves, creep parameters as well as mechanical strength properties of the tested PE1000 material for various stress levels and temperatures were obtained. A temperature increase degrades the strength and creep parameters of the material.

Introduction

The material investigated in the paper is polyethylene UHMW (ultra-high molecular weight) – a polyolefin type plastic material i.e., polymer containing only carbon and hydrogen atoms [1]. It is characterised with a number of excellent performance properties such as significantly high resistance to friction wear and perfect sliding properties. These properties are utilised to produce various machine and equipment components, such as bars, sliding guides used on factory production lines, gears, chain wheels, etc [2], [3]. PE-UHMW is also used as an additional cover material to protect surfaces from abrasion, therefore, it is used in cargo spaces in cars or railway wagons. PE-UHMW with its high impact resistance is used in production of various elements operated at low temperatures and exposed to impact loads, for example, fenders, bands and others. It has a number of applications in fuel-energy, electromechanical, textile and armament industries. It is worth mentioning that this material is inert in contact with a human tissue, which allows utilising it in the medical industry [4] and [5].

Plastic materials combine the characteristics of an elastic body and viscous liquid, which is called viscoelasticity. While examining the rheological phenomena in viscoelastic materials, the basic role is attributed to a time factor that determines the tested material behaviour under certain stress conditions. Rheological phenomena include creep and relaxation.

The structure of plastics consists of individual chemical compounds - monomers that are combined into long chains to form a polymer. Creep results from the chains straightening, their relative to each other movement and orientation in the acting force direction. [6]. The main

Experimental Mechanics of Solids   Materials Research Forum LLC
Materials Research Proceedings **12** (2019) 9-18   https://doi.org/10.21741/9781644900215-2

factor initiating the occurrence of the rheological phenomena is (apart from the load) the temperature.

A wide range of the material applications, including elements exposed to long-term loads and working at elevated temperatures, lead to conduct rheological tests. Moreover, polymer materials tend to deform even at low temperatures and under the influence of seemingly small loads. This phenomenon is not desirable and has a profound effect on the usability of an element made of such a material. Therefore, a plastic creep phenomenon needs to be taken into account at the product design stage. The material is usually selected and engineered to operate at a specific temperature and at a specified permanent load, therefore, the creep phenomenon can occur in a limited and predictable manner. In this case, a creep curve is similar to a straight line parallel to the timeline [7], [8]. This simplified approach can be insufficient in more demanding applications.

Proper multiple-stepped creep tests are generally time consuming, expensive and should be carefully planned. A 24 h time frame of creep test, presented in the paper, can be a reasonable option for practical applications and sufficient for numerical modelling purposes. In a few papers, the authors conducted creep research in time limited to 24 h.

In the PhD thesis [9], time-dependent constitutive relationships for various polymers polyethylene materials were carried out. The experimental results indicated strong nonlinear viscoelasticity in the material responses. Samples of seven polyethylene materials were tested. Six of them were high density polyethylene (HDPE) and one was medium density polyethylene (MDPE). 24-hour creep tests on seven materials were conducted for modelling purposes. Moreover, multiple-stepped creep tests were also performed. All the tests were conducted at room temperature.

It was observed that at high stress levels a creep strain grows rapidly from the beginning of the test and a large deformation occurs within the first two hours. An excessive deformation, far beyond the practically acceptable levels, was observed. A hardening plateau was observed closer to the end of the tests.

In paper [10], the authors developed a method for constitutive nonlinear viscoelastic modelling of polyethylene utilizing both the multi-Kelvin element theory and the power law functions to model a creep compliance. Creep tests were used to determine material parameters and models. Four different PE materials were tested in creep for 24 h. Three of the materials were specified as HDPE and one as MDPE. Five stress levels were set.

Experimental research

The UHMW (ultra-high molecular weight) type polyethylene PE1000, in the form of a 4 mm thick plate was tested experimentally. 24 h tensile creep tests, preceded by standard tensile tests, were performed for various temperatures and different stress levels. Flat samples were cut out using a water jet machine (Fig. 1). Two types of specimens were developed based on ISO-527 standard [11]. Specimens in the shape of 1B type were used in the static test at an ambient temperature (Fig. 1a). The length of the parallel measuring part of 1B type specimen was equal to 60 mm and the cross-section dimensions were equal to 10x4 mm. Specimens in the shape of 1BA type were used in the static test at elevated temperatures and in all creep tests (Fig. 1b). They had a shorter measuring part, due to chamber height limitation. The corresponding dimensions in the case of 1BA specimen were equal to 20 mm and 5x4 mm.

Experimental Mechanics of Solids
Materials Research Proceedings **12** (2019) 9-18

Materials Research Forum LLC
https://doi.org/10.21741/9781644900215-2

Fig. 1. ISO-527 specimens: a) 1B type for static standard test; b) 1BA type with a shorter measuring length for the tests in the temperature chamber

The structure of plastic exhibits a load memory effect [12]. Thus, the tested samples were pre-conditioned [13]. Before the tests, the samples were kept at the desired temperature for 4 hours or longer. Also, a furnace with the jaws inside, installed in the testing machine, was set to the desired temperature for at least 12 h to assure a uniform heat distribution in the entire work space of the chamber and in the jaws.

The PE plastic is characterized by a low moisture absorption equal to <0.01% [3], [14]. Therefore, it was assumed that the humidity in the test bench room has an insignificant effect on the material properties.

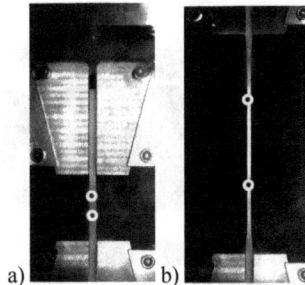

Fig. 2. An exemplary specimen installed in the temperature chamber: a) at the beginning and b) at the end of the test, with elongation limited by the furnace chamber dimensions

All the tests were performed on a creep testing machine equipped with an integrated videoextensometer. The videoextensometer allowed for elongation measurement of the specimen measuring length, excluding gripping parts, using digital image correlation methods (DIC). An exemplary specimen installed on the testing machine in the temperature chamber, with round black-and-white measuring marks for strain measurement, is presented in Fig. 2. Testing with a larger deformation is restrained in this study by instrumental limitations. Despite a shorter length of 1BA specimens, the internal dimensions of the temperature chamber limited the traverse displacement range and, thus, the maximum elongation of a sample during all tests at elevated temperatures (Fig. 2b) to 132 mm.

A test program was established using three creep constitutive function variables: temperature, load and time. The maximum duration of the tests was set to 24 hours. The temperature values were determined based on the material PE1000 characteristics. Three temperature values were chosen: 22°C – corresponding to the normative test temperature [13]; 42°C – the tested material melting point, according to the Vicat method A [5], [15] and 90°C – marked as the maximum long-term working temperature for PE1000 polyethylene. Furthermore, the functional considerations of the material, in terms of a thermal point, (e.g. contact with human body or boiling water) were also taken into account.

Static test results

To begin the creep experiment of the PE1000 material, it was necessary to perform static tensile tests. The material properties designated in the tensile tests provided a general view of the polyethylene UHMW behaviour. All static tests were carried out in accordance with the standard procedure [11] in the range of three fixed temperatures. Six specimens were tested for each temperature level. 1B type specimen shape was used at an ambient temperature (Fig. 1a), whereas 1BA type was used at elevated temperatures (Fig. 1b).

Static tests at 22°C resulted in the fracture of all samples. The measuring section of the samples deformed permanently along its entire length (Fig. 3). There was no clear necking of the measuring section. The material flowed evenly, constantly reducing the cross-sectional area of the section over time. The fracture formed edges almost perfectly perpendicular to the sample longitudinal axis. In Fig. 3, numerous buckling and torsions can be observed, which indicates a significant share of elastic energy in the sample prior the fracture.

The static tests at elevated temperatures lasted until the specified specimen elongation, limited by the size of the furnace chamber, occurred. The samples had a relatively uniform narrowing of the measuring section along its entire length.

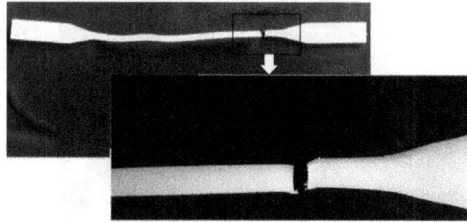

Fig. 3. Fracture of the exemplary specimen after the static test at an ambient temperature

Table 1. The results of PE1000 static tensile test at ambient and elevated temperatures

Material property description	Units	Temperature [°C]		
		22	**42**	**90**
Young modulus E	[MPa]	1132.0	680.5	206.6
Standard deviation σ_E		10.7	59.3	19.4
Percentage decrease	[%]	-	39.9	81.7
Ultimate strength R_m/ Stress at yield R_e	[MPa]	21.3	15.3	7.6
Standard deviation σ_{Rm}		0.43	0.36	0.17
Percentage decrease	[%]	-	28.2	64.3
Strain corresponding to ultimate strength ε_{Rm}	[%]	13.0	53.2	61.1
Standard deviation		2.20	2.40	7.30

The results of polyethylene PE1000 static tensile tests at ambient and elevated temperatures are presented in Table 2. The elementary material properties such as Young modulus, ultimate strength and strain corresponding to ultimate strength were obtained for each given temperature

Experimental Mechanics of Solids
Materials Research Proceedings **12** (2019) 9-18

Materials Research Forum LLC
https://doi.org/10.21741/9781644900215-2

level. Stress at yield point for typical plastic material curve according to [11], as in the case of the PE1000, has the same value as the ultimate strength. Standard deviation values were calculated. Also, a percentage decrease in the obtained material properties, relative to those achieved at ambient temperature, was determined.

σ-ε tensile curves of PE1000 plastic for each temperature level based on the median values are presented in Fig. 4. It can be noticed that with an increase in the temperature of the material, the curve shape changes in the area of the yield point. The curve begins to take the form similar to a waveform with a conventional yielding point [11] and with postponed ultimate strength. The elastic range of the material response is significantly reduced. In the tests at 90°C and 42°C, the samples fractures were not achieved due to the temperature chamber size limitations. Hence, their σ–ε curves are shorter compared to the test at 22°C.

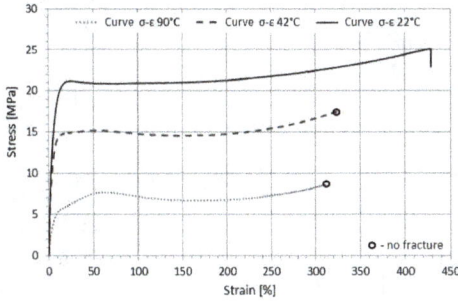

Fig. 4. Tensile curves of PE1000 plastic for each temperature level

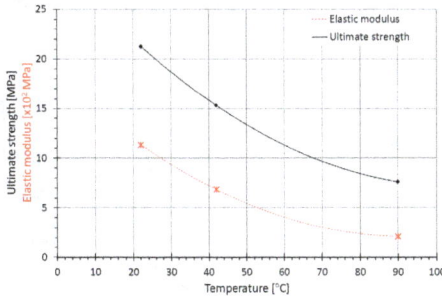

Fig. 5. Degradation of the PE1000 performance with temperature growth

Based on the presented results, it can be observed that the temperature rise degrades the material strength properties. With growth of the temperature, tensile curve histories (Fig. 4) as well as the material properties (Table 1) decrease. The dependencies of material properties on a temperature growth are nonlinear. In Fig. 5, tensile strength and elastic modulus data were approximated with polynomials of the second degree. The tensile strength value is reduced from 21.3 MPa, at an ambient temperature, to 15.3 MPa at 42°C and 7.6°C (64.3% decrease) at the highest temperature level. The elastic modulus declined from 1132 MPa to 680.5 and even to

203.6 MPa with respect to the temperature. The material stiffness is reduced by 81.7% at 90°C, compared to properties at an ambient temperature (Table 1).

Creep test results

Creep tests were conducted at 22°C, 42°C and 90°C on the creep testing machine presented in Fig.2. 1BA modified sample type was used in the experiment. The tests were carried out on the basis of a procedure compliant with the PN-EN ISO 899-1 standard [16]. A creep test program had time limitations. However, to obtain statistical data in the selected cases three specimens were tested. The most representative data are shown in the paper. It was found that at the beginning and in the middle phase of creep, the specimens behaved repetitively. Some differences in creep curves occurred at the end of the tests.

The average material strength of the PE1000 at a given temperature (Table 1) was a reference point for stress values at creep. After calculation in respect to a cross section area it gives a reference stress of 21.3 MPa (load of 420 N) at 22°C, 15.5 MPa (310 N) at 42°C and 7.5 MPa (150 N) at 90°C. The samples were loaded at given temperatures with seven multipliers of the reference stress: 0.9; 0.85; 0.8; 0.7; 0.6; 0.5; and 0.4. A stress level multiplier corresponds to a stress to ultimate tensile strength ratio. A detailed test program of creep loads is listed in Table 2.

Table 2. 24-h creep test programme of PE1000 plastic at ambient and elevated temperatures

Stress level multiplier (stress to ultimate tensile strength ratio)	Temperature					
	22°C	**42°C**	**90°C**	**22°C**	**42°C**	**90°C**
	Load at creep [N]			Stress at creep [MPa]		
Reference value for 1x multiplier	420	310	150	21.3	15.5	7.5
0.9	383	279	135	19.1	14.0	6.8
0.85	361	263	127	18.1	13.2	6.4
0.8	340	248	120	17.0	12.4	6.0
0.7	298	217	105	14.9	10.9	5.3
0.6	255	186	90	12.8	9.3	4.5
0.5	213	155	75	10.6	7.8	3.8
0.4	170	124	60	8.5	6.2	3.0

The tests lasted 24 hours at maximum which gives 86 400 seconds. Time reduction to 24 hrs may be a reasonable option for practical applications and sufficient for numerical modelling purposes,[9], [10].

The test was usually finished after a 24h period. Some load levels were so high at a given temperature that the creep test was completed in a much shorter time than set. At 22°C, such cases resulted from the samples breaking. Four of the seven samples (for multipliers of 0.9, 0.85, 0.8 and 0.7) fractured, while three with the lowest multipliers lasted for 24 hours. The pictures of the samples after the creep test are presented in Fig. 6a). The specimens in Fig. 6 were placed next to each other, in order from the most loaded (on the left side) to the least loaded (on the right side).

At a higher temperature level, none of the samples was fractured and, in this case, reaching the safety traverse limit ended the tests earlier. This was the case for stress multipliers of 0.9,

0.85 and 0.8. The other specimens lasted for 24 hours. The pictures of the samples after the creep at 42°C are presented in Fig. 6b). The corresponding 24h-creep curves are presented in Fig. 8.

In the case of the tests at the 90°C, safety limits were not also obtained. The pictures of the samples are presented in Fig. 6c). 24h-creep curves at 90°C are presented in Fig. 9.

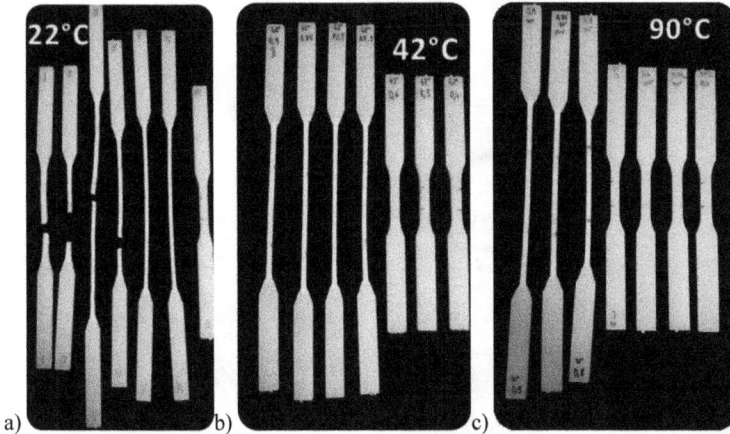

Fig. 6. Specimens after the 24h creep tests at ambient temperature (a) at 42°C (b) and at 90°C (c). Stress level decreases from the left to the right side of the pictures

As presented in Fig. 6a-c), fractures of the broken specimens were similar to those formed in the static tests. The samples which were not broken had a relatively uniformly narrowed measuring section along its whole length.

The final elongation depends on the stress level. The lower the stress the less the total elongation. This is the case for specimens stressed at 42°C and 90°C. At those temperatures none of the specimens fractured. The specimens stressed with a low multiplier (to 0.6 at 42°C and 0.7 at 90°C) obtained significantly lower total elongation than the specimens with higher multiplier, in which participation of ductile strain is high. Minor elongation differences can be caused by differences in fixturing of a given specimen.

At an ambient temperature, the resulted elongations of broken specimens are inconsistent. One of the specimens (third from the left in Fig. 6a) stressed to a lower level was subjected to a more ductile deformation in the total elongation than other fractured specimens. Moreover, in the case of the specimens loaded with the highest stress multipliers (0.9 and 0.85), the ductile strain share was lower compared to others, except for the 0.4. multiplier.

24h-creep curves of PE1000 plastic at 22°C are presented in Fig. 7. At ambient temperature, the material passes, during creep, into a state of plastic flow under loads equivalent to, or more than, $0.5R_{(22°C)}$, which is about $\sigma=10$ MPa. At 42°C (Fig. 8), the second creep stage occurs under stress equal to at least $0.7R_{(42°C)}$ ($\sigma=10.0$ MPa). At 90°C (Fig. 9), this is the case under stress of $\sigma=6.0$ MPa and more ($\geq 0.8R_{m(90°C)}$). The higher the temperature of the material, the lower stress is able to induce the flow process, however, with respect to the strength at a given temperature, a creep stress to an ultimate strength ratio increases.

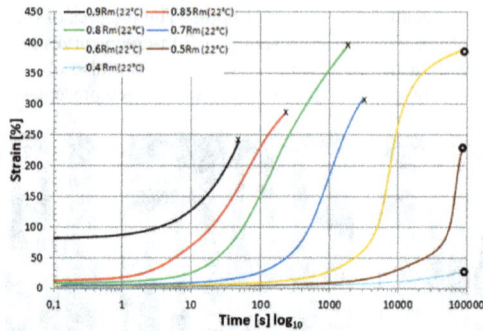

Fig. 7. 24h creep curves of PE1000 plastic at 22°C

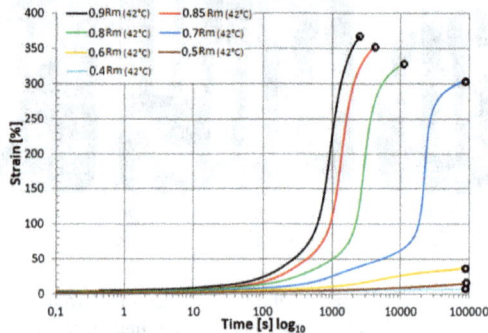

Fig. 8. 24h creep curves of PE1000 plastic at 42°C

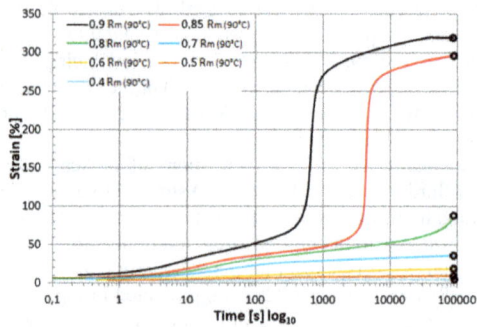

Fig. 9. 24h creep curves of PE1000 plastic at 90°C

At 22°C an unexpected phenomenon occurred in the case of creep curves at stress level multipliers of 0.5 and 0.6. After about 1000 s or more, the curves rapidly increased as if it would be III creep stage and, after reaching a specific higher strain, a strain ratio decreased (Fig. 7). For higher multipliers, the specimens crushed. This phenomenon is explained by hardening of the material under high deformations, which caused a strain ratio to decrease. A similar phenomenon is present in the case of the elevated temperatures, at 42°C, in the case of multipliers of 0.7 and higher and at 90°C, in the test of a 0.85 and a 0.9 stress multiplier. In [10] a similar behaviour is present in some cases of 24 h-creep graphs for polyethylene, however, it is not explained.

Table 3. Time required to achieve specific deformation of 30% and 150% or crush

Temperature	22°C			42°C		90	
Reference stress value	21.3			15.5		7.5	
Strain level	30%	150%	crush	30%	150%	30%	150%
Stress level multiplier	Time [s]						
0.9	–	16	48	136	781	10	636
0.85	2.8	41	226	211	1187	37	4673
0.8	13	98	1786	415	2655	80	–
0.7	120	778	3035	1331	21389	1738	–
0.6	1149	6436	–	18233	–	–	–
0.5	9582	67156	–	–	–	–	–
0.4	–	–	–	–	–	–	–

The temperature increase degrades the strength and creep parameters of the material. During creep, both at an ambient temperature and at the elevated temperatures, the dependence of the time required to achieve a specific plastic deformation or crush, as a function of the nominal stress, decreases exponentially. In Table 3, times required to achieve a specific deformation or crush are listed. Only specimens evaluated in 22°C were broken. For 0.9, crush occurred after 48 s while in the case of 0.8 –after 1789 s and 0.7 – after 30355 s. Strains of 30% and 150% were arbitrary selected as an example. 150% of strain at 42°C is obtained after 781 s for 0.9 multiplier and after 21389 s for 0.7 multiplier. Similarly, at 90°C, 150% is reached after 636 s at 0.9 load level and 4673 at a 0.85 load level (Table 3).

Conclusions
The final conclusions on the basis of the static tensile and 24h-creep tests under the tensile load of polyethylene PE1000, at ambient and elevated temperatures up to 90°C, were formulated:
- The temperature increase degrades nonlinearly performance of the material. The strength properties such as ultimate tensile strength and Young's modulus are reduced;
- At an ambient temperature, the material passes, during creep, into a state of plastic flow under stress values: $\geq 0.5R_{(22°C)}$, $\geq 0.7R_{m(42°C)}$ and $\geq 0.8R_{m(90°C)}$ at corresponding temperature. The higher the temperature of the material, the lower stress is able to induce the flow process, however, with respect to the strength at a given temperature, a creep stress to an ultimate strength ratio increases;
- The dependence of the time required to achieve a specific plastic deformation or crush, as a function of the stress level at given temperature, decreases exponentially;
- Under high load level multipliers, at deep II creep stage, the curves rapidly increase as if it would be III creep stage and, after reaching a specific higher strain value, a strain ratio decreased. This phenomenon is explained by hardening of the material under high deformations, which caused a strain ratio to decrease.

Experimental Mechanics of Solids   Materials Research Forum LLC
Materials Research Proceedings **12** (2019) 9-18   https://doi.org/10.21741/9781644900215-2

References

[1] S. Ochelski, Experimental methods in construction composites mechanics [in Polish], WNT Wydawnictwa Naukowo-Techniczne, Warsaw 2004

[2] M. Kutz, Applied plastics engineering handbook. Processing and materials, Elsevier, Waltham (2011)

[3] Information on: https://www.redwoodplastics.com/wp-content/uploads/2016/02/Redco-UHMW-Performance-Plastics-Feb2016.pdf

[4] Information on: https://www.mtf.stuba.sk/buxus/docs/internetovy_casopis/2010/3/szeteiova.pdf

[5] Information on http://www.plastics.pl/content/pliki/213/katalog_tworzywa_tech17.pdf (2018)

[6] D.V. Rosato, M.G. Rosato, D.V. Rosato, Concise encyclopedia of plastics, Springer-Science+Business Media, LLC, Kluwer Academic Publishers, Boston (2000). https://doi.org/10.1007/978-1-4615-4579-8

[7] M. Ashby, H. Shercliff, D. Cebon, Material engineering vol.1 [in Polish], Galaktyka, Łódź (2010)

[8] A. Jakowluk, Creep and fatigue processes in materials [in Polish], WNT, Warsaw (1993)

[9] L. Hongtao, M.A. Polak, A. Penlidis, A Practical Approach to Modeling Time-Dependent Nonlinear Creep Behavior of Polyethylene for Structural Applications, Polymenr engineering and science (2008), pp.159-167. https://doi.org/10.1002/pen.20942

[10] L. Hongtao, Material Modelling for Structural Analysis of Polyethylene A thesis of Master of Applied Science, University of Waterloo, Waterloo, Ontario, Canada (2007)

[11] PN-EN ISO 527-2:2012 Plastics -- Determination of tensile properties -- Part 2: Test conditions for moulding and extrusion plastics [in Polish], Polski Komitet Normalizacyjny PKN, Warsaw (2012). https://doi.org/10.3403/30216860

[12] I. Hyla, Plastics – properties – processing – application[in Polish], Wydawnictwo Politechniki Śląskiej, Gliwice (2004)

[13] PN–EN ISO 291:2010 Plastics -- Standard atmospheres for conditioning and testing [in Polish], Polski Komitet Normalizacyjny PKN, Warsaw (2010)

[14] Information on https://crownplastics.com (2018)

[15] S.M. Kurtz, The UHMWPE Handbook: Ultra-High Molecular Weight Polyethylene in Total Joint Replacement, Elsevier, Philadelphia (2004)

[16] PN–EN ISO 899-1:2005 Plastics — Determination of creep behaviour —Part 1:Tensile creep [in Polish], Polski Komitet Normalizacyjny PKN, Warsaw (2005)

Experimental Mechanics of Solids
Materials Research Proceedings **12** (2019) 19-30

Materials Research Forum LLC
https://doi.org/10.21741/9781644900215-3

Propagation of Failure in Adhesive Joints between CFRP/GFRP Beams Subjected to Cyclic Bending

Yuliia Terebus[a], Dominik Głowacki[b*], Mirosław Rodzewicz[c]

Warsaw University of Technology, Institute of Aeronautics and Applied Mechanics,
Nowowiejska 24 00-665 Warsaw, Poland

[a]yuliia.terebus@gmail.com, [b]dglowacki@meil.pw.edu.pl, [c]miro@meil.pw.edu.pl

Keywords: Composite Material, Adhesive Joint, Fatigue Test

Abstract. Last decades were crucial for composite structures. They were highly developed and started to be widely used in aircraft industry. This allowed engineers to overcome a set of issues connected with incapability of known materials to support modern requirements and needs. Merging together two or more materials in special way enabled to save advantages of each and get rid of individual drawbacks. As a result such a "hybrid" material got improved structural properties. However, in the aircraft industry important are not only individual parameters of the material, but how it can be joined with other part. Mechanical fasteners and adhesives types of joining are used with composites materials depending on application and material composition. Even as both types are still used, adhesive bonding of composite material are more beneficial in comparison with traditional mechanical fasteners. The *aim* of this article is to perform fatigue test of adhesive joints between CFRP and GFRP composites and analyze obtained results. In order to do that, destructive loads levels were investigated with mathematical model and with FEM simulation in ANSYS software.

Introduction

Reliability and safety play a primary role in the functioning and development of aviation. Since the beginning of aviation progress the reliability of the aircraft structure has been very important. That is why improving the quality, reliability and durability of structural elements have a special place in research investigation.

Nevertheless, the greatest interest have composite materials which are widely used in the aircraft industry and today can be found in many different kinds of airplanes, as well as gliders. Aircraft structures are commonly made up of 50 to 70 percent composite material. Moreover, it is hard to predict and detect the material fracture. So, reliability of composite structures plays an important role in aviation industry [1]. Very often a critical element regarding reliability is a joint between structural elements. The paper concerns investigations into adhesive joint (glue-connection) between CFRP and GFRP flat shells. Several methods are in use for testing such a joints [2, 3], but for the purpose of the research presented in this paper there were elaborated a special method. The uniqueness of this method lies on generating the state of shear stresses in adhesive joint between two composite shells glued together, by subjecting them into bending with constant value radius curvature. There were applied the specimen in the form of rectangular beams consisted of two halves: CFRP and GFRP orthogonal laminates, which were glued together only in a certain parts of their surface. In that way there were obtained the conditions of "notched" connection, which facilities the debonding between both halves, and effectively increases it's grow (especially in a low-cycle fatigue mode).

This aspect was very important for the authors, because this research was considered by them as a "pre-check" of a new testing method, so the results and answer regarding applicability of this method should be delivered in a short time. It is also worth to mention that the same specimens were used in another research (i.e. detection of debonding zone using ultrasonic flow detector), which is not described in this paper. That is why there were applied the specimens consisted of transparent and non-transparent composites, as the progress of debonding can be easily observed from "transparent side" of the specimen (and be measured using the caliper), so the results of those observations can be used for verification of ultrasonic inspection results, which were obtained from "non-transparent side". Nevertheless CFRP/GFRP structures appear to be used in practice, (e.g. in AOS-71 glider structure [4]).

Specimen Preparation
Composite specimens were manufactured at Warsaw University of Technology. They were cut from CFRP/GFRP workpiece (i.e. composite plate made from carbon and glass fabrics, and epoxy resin). Materials used for the workpiece fabrication were: Interglas 92125 (280 g/m² glass fabric), ECC 452 (204 g/m² carbon fabric), Epidian 53/Z1 (resin/hardener mixed proportion: 100:10.5, volume fracture: 0.5). A following layout was applied: 5 layers of carbon fabric consequently and then 5 layers of glass fabric; in each layer the reinforcement fibers were oriented alongside the plate edges.

The plate had an inclusion in the middle of its thickness (i.e. between CFRP and GFRP composite) made from 0.02 mm polyester foil (see Fig. 1). The role of this foil was to shape the dimensions of adhesive joint between CFRP and GFRP halves of the specimen and to simulate initial failure of this connection. Therefore, the foil was shaped as it is shown on the schematic sketch in Fig. 1.

Manufacturing Procedure:
1. Two flat pieces of laminated wooden board were prepared as molds.
2. The carbon fabric and glass fabric pieces of necessary dimensions were cut out from the roll. The piece of polyester foil with cutouts was prepared.
3. Molds and polyester foil were coated

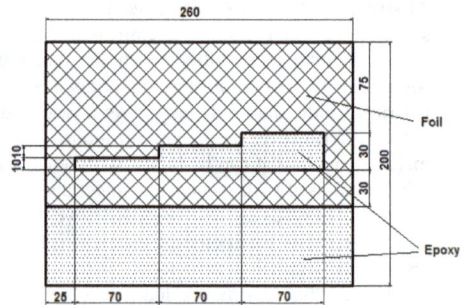

Fig. 1. Schematic sketch of the middle-plane cross-section of composite workpiece

with wax to provide easy separation of the workpiece from mold.
4. The sheets of reinforcement fabrics were consequently put on the mold and impregnated by Epidian 53/Z1 resin mixture (at first 5 carbon, then the separation foil, and then 5 glass). The composite plate was hand-rolled after impregnation of each layer to remove air

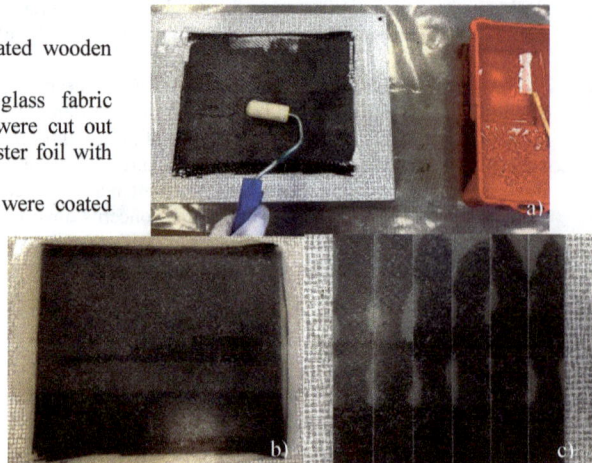

Fig. 2. Specimen manufacturing:
a) layers impregnation with epoxy;
b) workpiece; c) specimens

20

bubbles and provide uniform distribution of epoxy (Fig. 2a).

5.   Second mold plate was put on the top of composite plate and pressed (about 100 kPa) during first step of 2-stages curing process (16 hours at room temperature and then 8 hours at temperature of 60°C).

6.   Finally the composite workpiece (see Fig. 2b) was cut into 6 rectangular specimens 150mm x 25mm (i.e. two specimens for each length of adhesive joints between CFRP and GFRP halves). As a result were obtained 3 pairs of specimens having 10, 20, and 30 mm adhesive joints to be subjected into fatigue tests (Fig. 2c).

Fatigue Test

Fatigue tests were performed in laboratory of the Faculty of Power and Aeronautical Engineering of Warsaw University of Technology.

There was used pneumatic actuator powered by pressure system consisting of long pneumatic piston-pomp and a screw-servomotor linked with the piston of this pomp (see Fig. 3). The servomotor was controlled by computer using input file with programed load/deflection sequence (i.e. sequence of pomp piston movement).

The base of the specimen was fixed in a special holder (assuring the limitation of specimen deflection under load), while the tip of specimens was subjected by the actuator to transversal movement (low-frequency oscillations). The holder consisted of two wooden blocks having radius of contact-surface 150 mm. In consequence the specimen was bended and the curvature of deflection was limited by the radius of those blocks (see Fig. 4).

The number of bending cycles applied was captured by electronic counter.

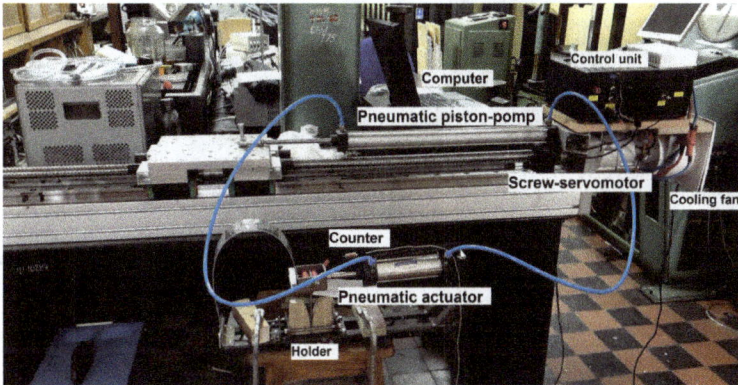

Fig. 3. Fatigue test stand

Performed experiment was a low-cycle fatigue test. The load cycles applied during tests were of the same amplitude and frequency.

Testing Procedure. Each specimen was attached to fatigue stand and subjected into cyclic bending. After each 100 bending cycles the distance of failure propagation in adhesive joint between CFRP and GFRP halves of specimen was measured using caliper, which was made possible by transparence of the GFRP part of the specimen (see Fig.5).

Fig. 4. Deflection position of specimens

The view of specimens after first 100 and 400 bending cycles is shown in Fig. 5.

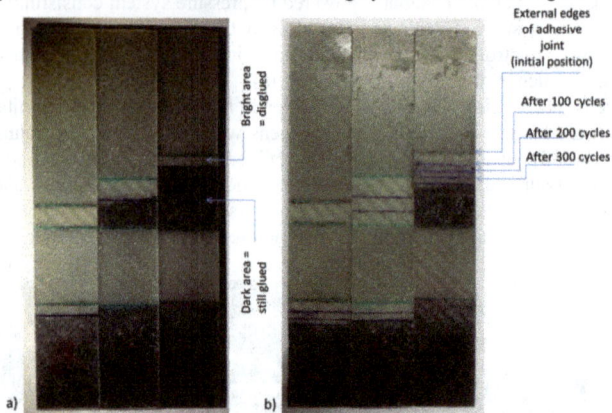

Fig.5 Specimens after 100 (a) and 400 (b) bending cycles

After finishing the set of experiments the following results were acquired:

- The specimen #2 (adhesive joint length 10 mm) has broked in 20 cycles;
- The specimen #4 (adhesive joint length 20 mm) has broked in 217 cycles;
- The specimen #6 (adhesive joint length 30 mm) has broked in 763 cycles.

They are collected in Tab. 1 and illustrated in Fig. 6.

Table 1. Fatigue test results

	Number of cycles	Failure propagation, [mm]
Specimen 2	0	0
	20	10
Specimen 4	0	0
	100	8.5
	200	14
	217	20
Specimen 6	0	0
	100	5
	200	8
	300	10.5
	400	13
	500	15.5
	600	18
	700	23
	763	30

Experimental Mechanics of Solids
Materials Research Proceedings **12** (2019) 19-30

Materials Research Forum LLC
https://doi.org/10.21741/9781644900215-3

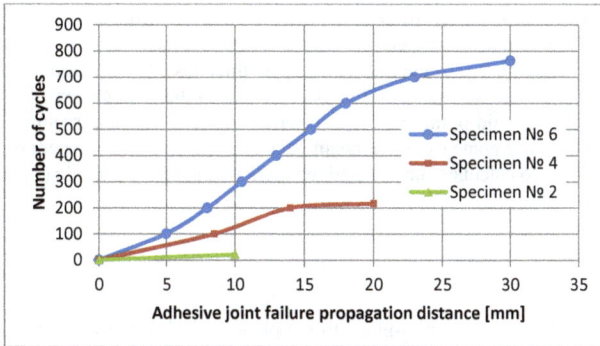

Fig. 6. Number of cycles to failure

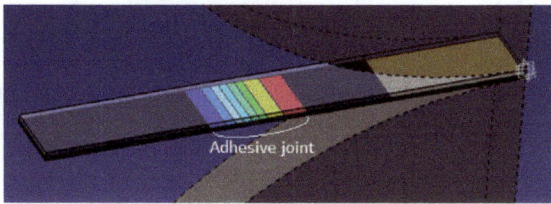

Fig. 7. Visualization of adhesive joint failure propagation

As it can be seen from the graph firstly failure propagates linearly up to last 10 mm. Only a few load cycles are needed to destroy last 10 mm. However, in specimen with adhesive joint lenght 30 mm the first 200 cycles led to higher failure propagation compared with next 500 cycles. It can be explained by high shear stress at the tips of adhesive joint in comparison to the stress in the middle part of this joint.

Estimation of Destructive Loads Level with Mathematical Model

Due to specimen cyclic bending the CFRP and GFRP parts of the specimen are compressed or stretched (depending on the direction of actuator movement). As a result the adhesive join between CFRP and GFRP halves of the specimen is subjected to cyclic shear deformation.

The load level in adhesive joint was estimated based on the simplified model, which is shown in Fig. 8. The thickness of adhesive layer was neglected here. At the beginning there were considered two theoretically possible situations: case "fully integrated", and case "without adhesion". For the first case the specimen has one neutral plane (i.e. plane with elongation equal zero), which is placed in the middle of specimen thickness near the contact

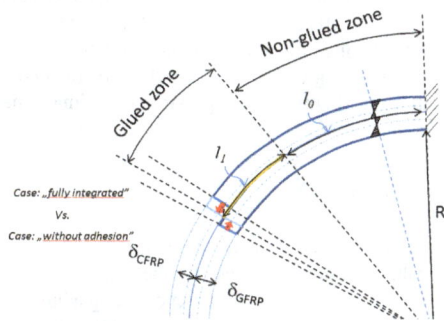

Fig. 8. Simplified model of the specimen for shear stress estimation in contact zone

23

Experimental Mechanics of Solids
Materials Research Proceedings **12** (2019) 19-30

Materials Research Forum LLC
https://doi.org/10.21741/9781644900215-3

zone of GFRP and CFRP halves (depending on the ratio of their bending stiffness). For the second case the GFRP and CFRP specimen-halves work separately as two bended beams. Each beam has neutral plane placed in the mid-point of its thickness. For a real case, one can expect the intermediate situation: i.e. the neutral planes of both bended halves have to be placed somewhere between the mid-point and contact zone. Due to different curvature of both bended beams the length of outer plane of GFRP beam becomes longer than the length of inner plane of CFRP beam. In order to calculate the elongations there was used the formula (1) concerning pure bending [5]:

$$\varepsilon = -\frac{z}{\rho},$$

(1)

where z – is a distance from the mid-plane to the plane considered, and ρ – is a radius of mid-plane curvature.

The relative elongations of outer plane of GFRP and inner plane of CFRP were calculated according formulas (2) and (3). The k-factor here is the ratio positioning neutral planes of bended beams between their inner and outer faces; (k=0,5 for the case "without adhesion").

In case when the specimen is deformed as it is shown in Fig.8, the strain of GFRP and CFRP can be calculated from equations (2) and (3) correspondingly:

$$\varepsilon_{GFRP} = \frac{k*\delta_{GFRP}}{R+\delta_{GFRP}} \text{ and } \varepsilon_{CFRP} = \frac{-(1-k)*\delta_{CFRP}}{R+\delta_{GFRP}+(1-k)*\delta_{CFRP}}$$

(2), (3)

Having values of those elongations and neglecting shear deformation of the glue,

it is possible to calculate axial forces necessary to compensate those elongations (see red arrows in Fig. 8) using formulas (3) and (4) , which are related to unitary width of the specimen.

$$F_{GFRP} = \varepsilon_{GFRP} * E_{GFRP} * \delta_{GFRP} * 1; \text{ and } F_{CFRP} = \varepsilon_{CFRP} * E_{CFRP} * \delta_{CFRP} * 1;$$

(3), (4)

Both forces should be equal for pure bending, and this fact depends on k-factor value.

In case of input-data used for considered model (i.e.: l_0=30mm, l_1= 10 or 20 or 30mm, E_{GFRP} = 20000 MPa, E_{CFRP} = 55000MPa, R=150mm, δ_{CFRP} =1,55mm, δ_{CFRP}=1,2 mm) – the balance was obtained for k-value equal 0,67629.

Assuming that in case of adhesive connection the same forces are produced by a shear stress in a glued zone, it was possible to estimate mean value of shear stress for each length of glue connection l_1:

$$\tau_{mean} = \frac{P_{GFRP}}{l_1*1}$$

(5)

The results of such estimations are stored in Table 2. It should be pointed, that maximum values of shear stresses in adhesive joint-ends may be higher because of hyperbolic-nature of shear-stress distribution function along adhesive joint [6].

Table 2. Shear stress estimation

Length of glue zone l_1	10	20	30
Shear stress [MPa]	13,8	6,9	4,6

Experimental Mechanics of Solids   Materials Research Forum LLC
Materials Research Proceedings **12** (2019) 19-30   https://doi.org/10.21741/9781644900215-3

As the shear stress limit for epoxy resin is about 30 MPa, it means that the loads in adhesive joint are relatively high and consistent with high-cycle fatigue testing idea.

Calculation of Destructive Loads Level with FEM Simulation

The next step of investigation into shear stress values at adhesive joint was implementation of finite element method (FEM) using ANSYS software system [7]. There was applied simplified 2D model (see Fig. 9). consisted from 2091 finite elements and 14761 nodes.

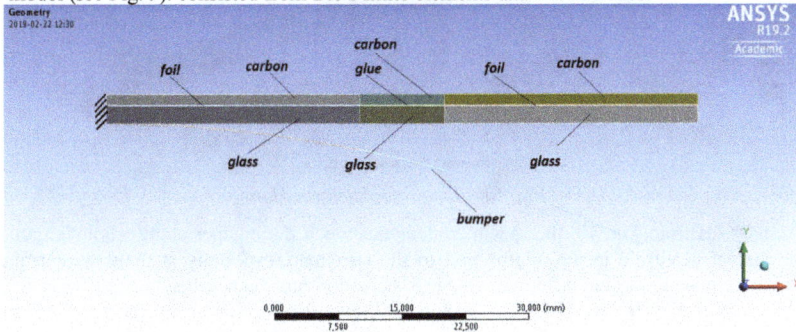

Fig.9. Structure of numerical model of the specimen

Geometry parameters of the specimen were the same like in previous chapter. There also were considered 3 different lengths of adhesive joint: 10, 20 and 30 mm. For all those cases the thickness of glue layer was the equal 0,1 mm. The places between specimen halves, which were outside the join were separated by the foil of the same thickness, but having very low E- and G-modulus in comparison with the material of adhesive joint.

For simplicity there was assumed that all specimen components were built from isotropic materials. Their parameters are specified in Table 3.

Table 3. Material parameters of FEM model

	Young's Modulus, [MPa]	Poisson's Ratio	Bulk Modulus, [MPa]	Shear Modulus, [MPa]
CFRP	54000	0.3	45000	20769
GFRP	20000	0.3	16667	7692.3
Distance foil	10	0.3	8.3333	3.8462
Adhesive joint (epoxy glue)	3000	0.3	2500	1153.8

The specimen-model was fixed at left end (see Fig. 9). On the opposite end, instead of load there was introduced the displacement along local Y-axis, which caused almost full contact of the specimen with model of "bumper" having curvature R=150mm. In order to avoid mutual penetration there were applied contact elements between the "bumper" and the specimen, which did not allowed deflection under "bumper" surface.

The results of calculations are presented below in the form of bitmaps.

The first one is a map of total deformation (Fig.10).

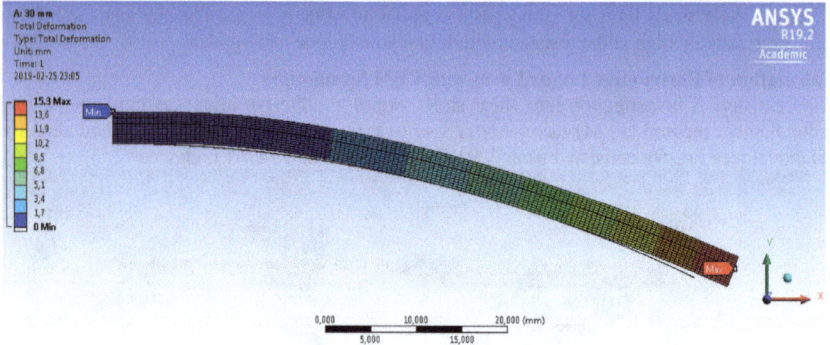

Fig. 10. Map of total deformation

As it arises from Fig. 10, the specimen does not touch the bumper along whole length, but good contact is visible in the middle part of the specimen, especially at the zone containing adhesive connection. The influence on such behavior has higher bending stiffness of the specimen in this zone. Variation of bending stiffness along specimen length generates shear stress concentrations in adhesive layer between CFRP and GFRP halves. In the real specimen the CFRP and CFRP halves can slide freely on the distance foil, while in analyzed, simplified model both halves are linked with the foil. There were not applied contact elements between foil and composite halves, but instead of it, there was assumed low E- and G-modulus. That is why one can observe a big shear

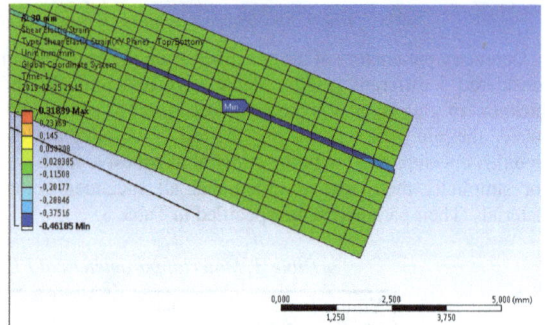

Fig. 11. Shear deformation the free end of the specimen

deformation of this foil at the free-end of the specimen (Fig. 11). This simplification of distance foil model should not affect significantly the shear stresses in adhesive joint. Illustration of shear stress distribution along the specimen is shown in Fig. 12. As it arises from this Figure – the maximum values of shear stress is about 40 MPa at the left tip of adhesive joint for all cases of adhesive length, and up to 14 MPa more at the right tip (depending on the case of adhesive length. In the middle part of adhesive joint shear stress is very low (see Fig.13).

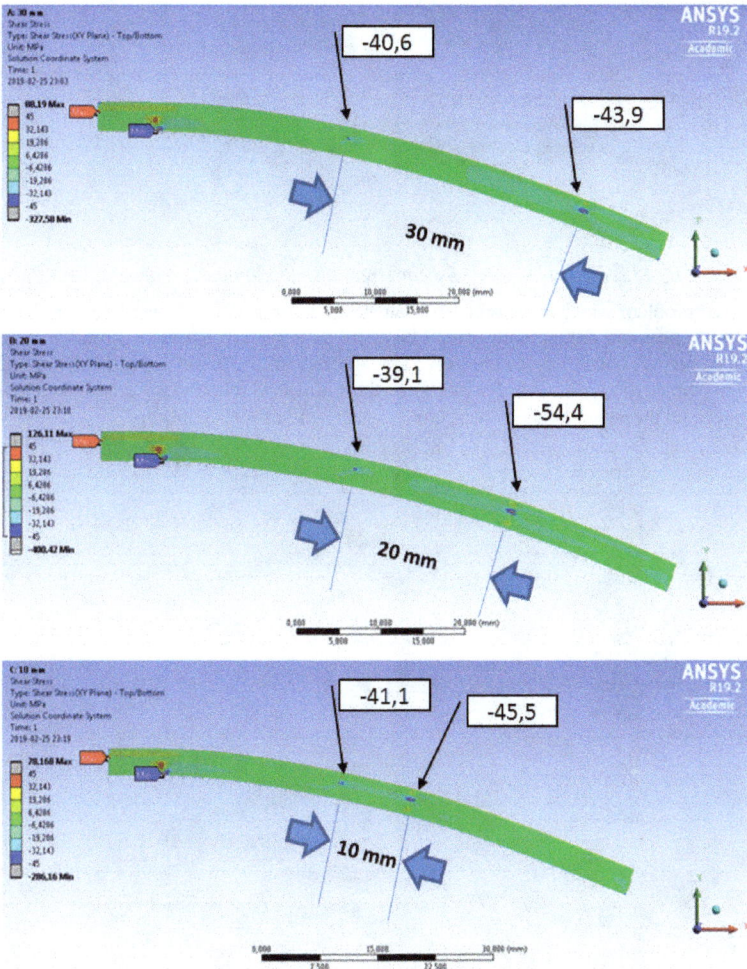

Fig. 12. Shear stress concentrations for different length of adhesive joint

Maximum values of the shear stress calculated from FEM-model exceed shear strength of epoxy-glue. From experimental part of work arises that the failure of adhesive joint did not appear during first load cycle, so it means that obtained values of shear stresses from FEM-model are overestimated. The reason for it is such that instead of multi-layer-orthotropic model of material there was used isotropic material.

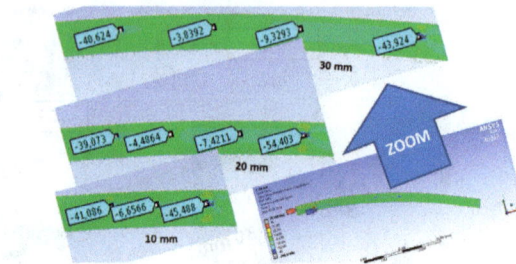

Fig. 13. Zoomed bitmaps of shear stress distribution inside adhesive joint

Another problem which is possible to evaluate basing on this FEM-model is the how much the state of specimen deformation using "bumper" as curvature template is similar to state of "pure bending". This question can be solved by analysis of distribution of normal stresses along bended specimen (Fig. 14). As it arises from the bitmaps – the stresses are uneven along the layer placed on the same radius of curvature, so such similarity is doubtful.

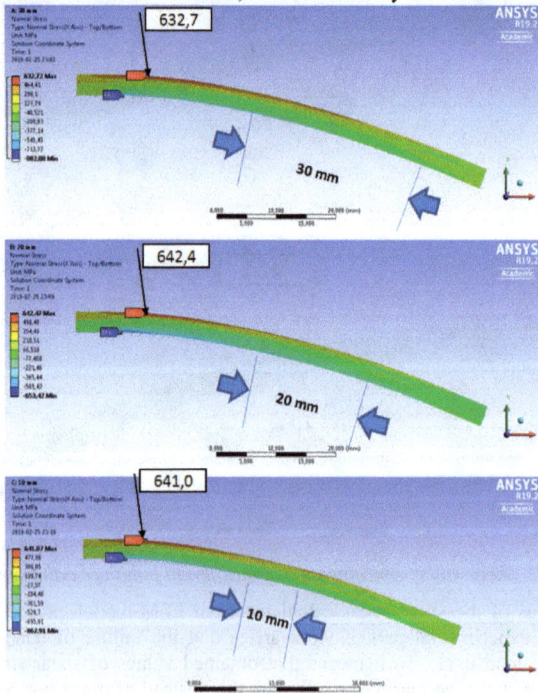

Fig. 14. Distribution of normal stress along the specimen

Investigation of Epoxy Fracture Type by Microscope Usage
To define how adhesive joint failed, the CFRP and GFRP specimen-halves were separated (see Fig. 15).

Fig. 15. CFRP and GFRP parts of the specimen (adhesive joint length 30 mm)

Photos of the CFRP and GFRP specimen-halves were taken using the digital microscope SMART SMP PRO. Photos were made of the inner side of specimen halves in non-glued and glued areas. This was done to understand in which way adhesive layer failed.

Several microscope photos were made and analyzed. The conclusion is that the fracture of adhesive joint between GFRP and CFRP halves of the specimen was cohesive, because the fragments of fractured epoxy layer are almost equally distributed on both halves of the specimen. Also are visible small de-cohesions inside GFRP material (see Fig. 16b), what proves good connection between GFRP and CFRP specimen-halves.

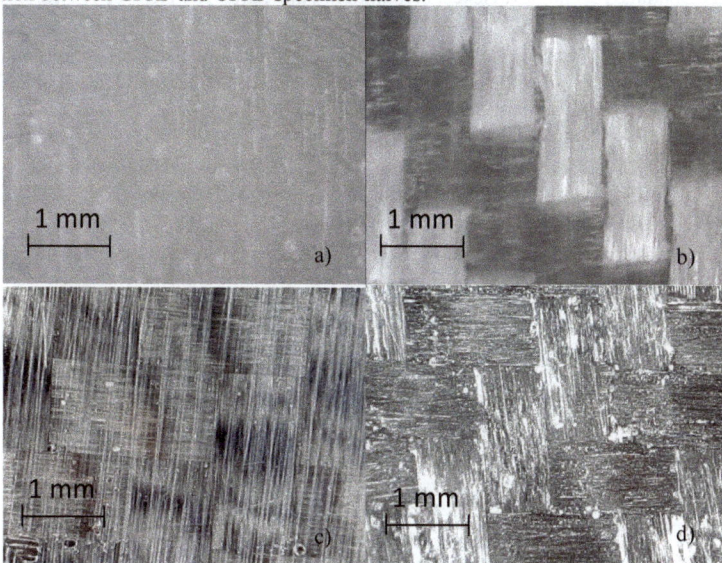

Fig. 16 Specimen areas under the microscope: a) GFRP surface non-glued; b) GFRP surface glued; c) CFRP surface non-glued; d) CFRP surface glued;

Summary
Low-cycle fatigue tests proved high performance of the adhesive joint between GFRP and CFRP halves of the specimens. Due to application of curvature templates with certain radius the specimen was subjected to bending. This bending induced shear loads inside adhesive joint between GFRP and CFRP parts. Three kinds of specimens with different length of adhesive joint

Experimental Mechanics of Solids
Materials Research Proceedings **12** (2019) 19-30

Materials Research Forum LLC
https://doi.org/10.21741/9781644900215-3

were tested. Nonlinear dependence of number of load cycles versus failure propagation was observed.

After set of experiments and analysis it was found that the length of adhesive joint has crucial influence on fatigue life. Increasing the length from 10 to 30 mm one can observe the change of the number of cycles to failure from 20 to 763.

The stress value at the end of adhesive joint was calculated using simplified mathematical model (based on deflected specimen geometry analysis) and it was also simulated using FEM calculations of 2D models with different length of the adhesive joint. Also FEM-model was simplified, because there were used isotropic materials instead of multi-layer-orthotropic model of composites, and relatively poor mesh.

Both numerical and FEM models allowed to estimate shear stresses acting on adhesive layer, but accuracy of this estimation is too low. The improvement of FEM-model should be considered as the future action.

References

[1] Composite Structures, Design, Safety and Innovation, Book 2005 Edited by B.F. Backman, Elsevier Science

[2] Redux Bonding Technology, Hexcel corporation, Publication No. RGU 034c – Rev, July 2003

[3] K. Machalická, M. Eliášová, Adhesive joints in glass structures: effects of various materials in the connection, thickness of the adhesive layer, and ageing, International Journal of Adhesion and Adhesives, Volume 72, January 2017, pp 10-22. https://doi.org/10.1016/j.ijadhadh.2016.09.007

[4] J. Marjanowski, J. Tomasiewicz, W. Frączek, The electric-powered motorglider AOS-71 - The study of development. Aircraft Engineering and Aerospace Technology. 89. 10.1108/AEAT-11-2016-0218. https://doi.org/10.1108/aeat-11-2016-0218

[5] Z. Brzoska, Wytrzymałość materiałów, Wyd. 4. Warszawa: Państwowe Wydawnictwo Naukowe,1983.

[6] IIT Kharagpur, Design of Adhesive Joints, Version 2 ME

[7] G. Krzesiński, T. Zagrajek, P. Marek, P. Borkowski, Metoda elementów skończonych w mechanice materiałów i konstrukcji: rozwiązywanie wybranych zagadnień za pomocą systemu ANSYS, Warszawa: Oficyna Wydawnicza Politechniki Warszawskiej, 2015.

Experimental Mechanics of Solids
Materials Research Proceedings **12** (2019) 31-36

Materials Research Forum LLC
https://doi.org/10.21741/9781644900215-4

Corrosive and Mechanical Experimental Tests for Selected Stainless Steel Pipes

Wasilewska Katarzyna[1, a *], Glinicka Aniela[2, b *]

[1]Warszawa, ul. Płużnicka 5c, Poland

[2] Warsaw University of Technology, Faculty of Civil Engineering, Warszawa, L. Kaczyńskiego 16, Poland

[a]wasilewska.katarzyna.1@gmail.com, [b]a.glinicka@il.pw.edu.pl

Keywords: Corrosion, Small Diameter Pipes, Stainless Steel, Experimental Studies

Abstract. The article presents corrosion tests of small diameters pipes made of stainless steel OH17N12M2 chromium - nickel - molybdenum with material number 1.4401 EN 10088 which does not have a pronounced yield point. The samples were subjected to corrosion in a laboratory at room temperature in solutions of sulfuric and hydrochloric acid at specified concentrations and in a given time. Corrosion was observed as surface uniform. As a result of the analysis, the dependences showing the mass loss as a function of time were obtained and the corrosion rate and average speed were determined. Following that, some of the corroded and non-corroded samples were subjected to a bending test in the Instron 3382 testing machine. On the basis of the measurements, plots showing maximum deflection of a beam as a function of bending load were created and compared. It was determined that the samples which were corroded for 720 h in 5% H_2SO_4 were not corroded and retained the bending capacity. Samples corroded for longer period of time or in more potent acids had the bending capacity reduced which was noted.

Introduction

Use of components made of stainless steel in normal atmospheric conditions does not cause the corrosion of steel they're made of. It's used in construction of bridge elements, buildings, hot water systems, sewage treatment plant elements, elements and constructions in chemical plants [1,2].

Stainless steel can corrode in the event of a failure, for example a sulfuric acid spill or under the conditions of constant operation in aggressive medium in the environment [3].

Let's assume that in a chemical plant there's a bearing construction made of small diameter steel pipes OH17N12M2; for example stairs, lattice, tower, support structure, etc. Let us ask questions that show the purpose of this work. How fast will stainless steel corrode in 5-20% H_2SO_4 and HCL acid after 720h or many times longer. What are the quantitative results of the loss of load bearing capacity when bending small diameter pipes made of stainless steel? Can corroded pipes still be used and if so, how much their use capacity have been reduced.

Corrosion of Stainless-Steel Samples

Samples accepted for testing were small diameter pipes made of steel OH17N12M2 chromium - nickel - molybdenum steel with material number 1.4401 EN 10088 in different diameters. The chemical composition as well as and physical and mechanical properties of this material are given in the standards tables [4, 5]. Standardized value of tensile strength of this steel is $R_m = 530 \div 680$ N/mm² and yield strength is $R_{02} = 240$ N/mm². A series of pipes with five different nominal diameters DN [mm] have been prepared, they were as follows: DN15, DN18, DN22, DN28 and DN35. There were 3 samples in each series.

Experimental Mechanics of Solids

Materials Research Forum LLC

Materials Research Proceedings **12** (2019) 31-36

https://doi.org/10.21741/9781644900215-4

The dimensions of series of tubular samples, designated for corrosion in sulfuric acid solutions were as follows (diameter Φ, wall thickness g, and length 1 given in [mm]): DN15: Φ15, g=1.2, *l*=300, *l*=30; DN18: Φ18, g=1.2, *l*=300, *l*=30; DN22: Φ22, g=1.2, *l*=300, *l*=30; DN28: Φ28, g=1.2, *l*=300, *l*=30; DN35: Φ35, g=1,5, *l*=300, *l*=30. The series of tubular samples prepared for corrosion in hydrochloric acid were as follows: DN15, DN18, DN22, DN28, DN35. Wall thickness was the same as in the samples above, length was $l = 30$mm. Corrosion tests were carried out for all samples and bending tests for 300mm pipes. Stress tests for pipes with length $l = 300$mm were carried out in the Instron 3382 testing. The corrosion process was carried out in the laboratory in acidic solutions, in cuvettes at ambient temperature (Tab 1) Corrosion time was set based on the literature (shortest time was 360h); a detailed description is available in the paper. Some of the cuvettes with corroding samples were placed under a vent and some were not. In the first case the acid concentration varied, but in the second case it was stable. Before the corrosion process, samples were weighed; after, they were dried and weighed again.

Table. 1. List of corrosion times and corrosive environments.

Corrosive environments	Corrosion times [h]
20% sulfuric acid, cuvette under a vent	720, 1228, 1704, 2208=tk
5 % sulfuric acid, cuvette under a vent	720
5 % sulfuric acid, open cuvette	720
20 % hydrochloric acid, open cuvette	720

The corrosion was assessed macroscopically as being uniform on surface on both inner and outer surfaces. After the measurements, the rate of the mass and the linear corrosion [3, 7] for each diameter of a pipe (Fig 1) were calculated. The average rate of corrosion was also calculated.

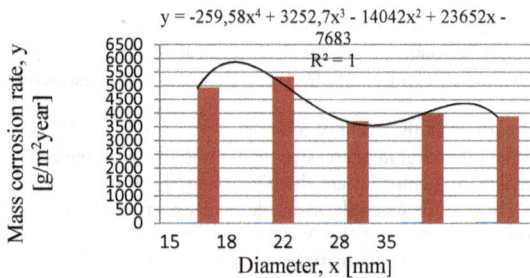

$$y = -259{,}58x^4 + 3252{,}7x^3 - 14042x^2 + 23652x - 7683$$
$$R^2 = 1$$

Fig. 1. Polynomial dependence, rate of corrosion – pipe diameter; 20% H_2SO_4

Average velocity of linear V corrosion, which does not vary with diameters, is: in 20% H_2SO_4 - $V= 0{,}53$ [mm/year], in 5% H_2SO_4 under the extract $V = 0{,}13$ [mm/year], in 20% HCL - $V=0{,}19$ [mm/year].

Statistical evaluation of the linear corrosion rate was carried out using Student's t-distribution [8]. Given $n = 20$ average values of linear corrosion rate measurements of stainless-steel samples corroded with 20% sulfuric acid, the mean value of $X = 0.53$ was obtained; and uncertainty of the mean value $S_{Xt} = 0{,}052$. Whereas, from the series of $n = 5$ average values of linear corrosion rates of stainless-steel samples corroded in 20% hydrochloric acid, we obtained: $X=0{,}19$; $S_{Xt}= 0{,}056$

Experimental Mechanics of Solids
Materials Research Proceedings **12** (2019) 31-36

Materials Research Forum LLC
https://doi.org/10.21741/9781644900215-4

The results of corrosion tests acquired after 720h lead to following conclusions:
1. The corrosion process in a 5% H_2SO_4 in an open cuvette has not happened, so it is a safe environment.
2. Corrosion process in 5% sulfuric acid solution under a vent happened, the vent caused evaporation of distilled water (even though the water was being added during the experiment) so the acid concentration could've changed. It is not a safe environment.
3. Corrosion processes were most intense in 20% sulfuric and hydrochloric acid solutions, so they are considered dangerous environments.
4. When corrosion time was longer, than appropriately longer than 720h, the loss of mass of the pipes was significantly higher. The amount of the loss was dependent on the acid concentration and length of corrosion.
5. Loss of mass shows non-linear dependency on the pipe's diameter. For applicability purposes, average values were calculated [9].

Bending Tests for Corroded Tube Samples
On the basis of axial tensile tests of non-corroded and corroded pipes in the Instron testing machine, it was determined that corrosion did not affect the Young's modulus E, yield stress threshold R_{o2}, elongation threshold u_{o2}, nor tensile strength R_m. It did decrease the elongation u_{om} and cross-section of samples [6].

This chapter presents the results of bending tests under static load performed in an Instron 3382 testing machine; Fig. 2 and Fig. 3. The samples were secured against bending via appropriate supports. The traverse speed of the machine was 5 mm/min and the frequency of data collection was 5 points/sec. Bending tests were performed on stainless steel pipes corroded in a 5% strength sulfuric acid under the extract for 720 hours, in a 20% strength sulfuric acid solution for 720 hours and in a 20% sulfuric acid solution for 2208 hours (tk - end time). Non-corroded pipes with the same dimensions were also bent. The measurements were made on pipes with diameters: DN22, DN28, DN35. The spacing of support beams was set to 250 mm. The measurements were carried out for 36 samples. This gives the opportunity to quantitatively assess the loss of bearing capacity [9, 10, 11]. Compression tests of short stainless-steel pipes corroded are described in [12].

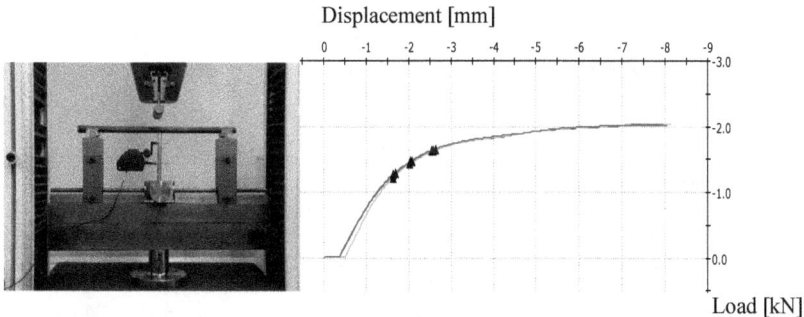

Fig. 2. Bending test of corroded pipe. Fig. 3. Load – displacement curves for 3 bending pipes DN22, which were corroded in 20% sulfuric acid for 720 h.

Experimental Mechanics of Solids   Materials Research Forum LLC
Materials Research Proceedings **12** (2019) 31-36   https://doi.org/10.21741/9781644900215-4

The software of the testing machine was called BlueHill [13] and was used to parse and compile the measurement results. On the plots depicting vertical displacement in the middle of a beam (deflection arrow) in relation to the load, which can be received with the BlueHill software, three characteristic points were selected for reading, namely: point 1 - corresponding to 33% of the load curve slope - deflection arrow, point 2 - corresponding to 50% of the tilt threshold, and point 3 - corresponding to 67% of the tilt threshold; Fig. 3. The load values and the deflection arrows were read from these graphs at the end of the linear section of the graph. The type of destruction of all the samples was the same, i.e. the first phase was characterized by a deformation of the upper wall followed by the second phase - the permanent destruction of the bottom edge. Fig. 4 plots show a relation between load and deflection arrow for tested beams. Based on the measured load values, the values of the relative decrease of the bearing capacity

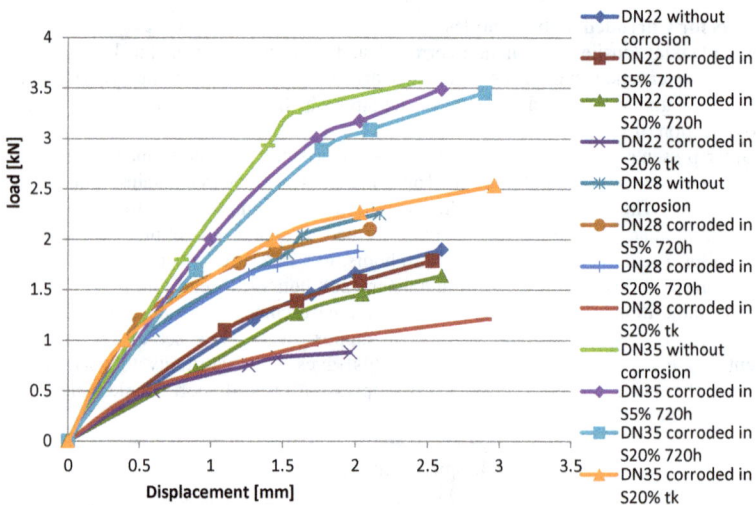

Fig. 4. Load - displacement graph for stainless steel pipes corroded in sulfuric acid and non-corroded.

transferred by corroded pipes in the points 1 to 3 and at the end of the linear section were calculated, and the results obtained are listed in tab. 2.

The drop of bearing capacity at the boundary of the linear section after corrosion of the pipe in 5% sulfuric acid for 720 hours was very low. The longer the sample corrosion time and the higher the sulfuric acid concentration, the lower the bearing capacity. The large value of bearing capacity decrease was recorded when stainless-steel pipes corroded for 2208 h (92 days) in 20% sulfuric acid. After the analysis carried out in [6], it was found that corrosion did not affect the Young's modulus E, tensile strength and the yield stress of the material. However, it decreased the cross-section of the samples.

The decrease equal zero of EJ_y occurred when the DN28 stainless steel pipes corroded for 30 days in 5% sulfuric acid and the DN22 pipes corroded for 30 days in 5% and in 20% sulfuric acid.

Table 2. Comparison of the relative load decrease in corroded bent pipes according to measurements.

Corrosion	Relative load drop at the end of the linear section [%]	Relative load drop point 1 [%]	Relative load drop point 2 [%]	Relative load drop point 3 [%]
DN22, sulfuric acid 5% 720 h	8	5	4	6
DN22, sulfuric acid 20% 720 h	42	14	12	14
DN22, sulfuric acid 20% tk	58	49	50	54
DN28, sulfuric acid 5% 720 h	-	5	8	7
DN28, sulfuric acid 20% 720 h	23	12	15	16
DN28, sulfuric acid 20% tk	54	49	48	46
DN35, sulfuric acid 5% 720 h	-	-	2	2
DN35, sulfuric acid 20% 720 h	6	1	5	3
DN35, sulfuric acid 20% tk	44	32	30	29

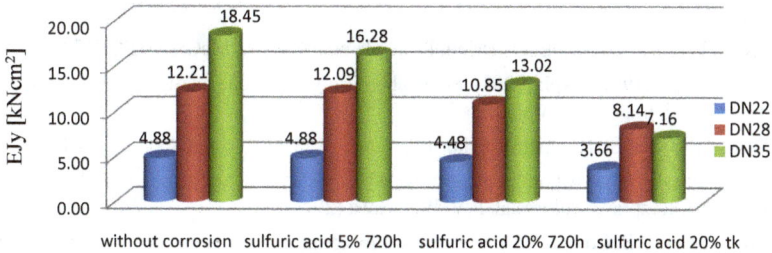

Fig. 5. The effect of corrosion on the EJ_y coefficient.

Conclusions

The summary presents the conclusions from corrosion tests and mechanical tests of stainless steel samples. To sum up analysis of the results of corrosion effects, carried out in the laboratory, on small-diameter pipes, it was found that:

1. corrosion that occurred was determined to be surface uniform; rate of corrosion is dependent on the exact diameter of the corroded pipe,
2. stainless steel is resistant to 5% sulfuric acid, but is not very resistant to 5% sulfuric acid under a vent (because its concentration may have increased), it is not resistant to 20% hydrochloric acid and 20% sulfuric acid. Because of this, only pipes made of stainless steel corroded in 5% sulfuric acid after 720 hours are suitable for use,
3. In the final effects, average corrosion rates have been calculated for corroded stainless steel pipes.
4. The results were statistically based on the Student's t distribution. The lowest uncertainty in the average value of the corrosion rate was obtained for corrosion of samples corroded in 20% sulfuric acid,

Experimental Mechanics of Solids   Materials Research Forum LLC
Materials Research Proceedings **12** (2019) 31-36   https://doi.org/10.21741/9781644900215-4

5. The most important factor impacting the loss of pipe mass was the type of corrosive environment and the time of contact of the pipes with this environment.

Summing up the analysis results of the bending tests of beams with circular cross-section, performed in the strength testing machine, it can be concluded that:

1. There were no differences in the shapes of damaged corroded and non-corroded pipes in bending tests,
2. The values of the drop of bearing capacity when bending corroded pipes were assessed,
3. Extending corrosion time caused an increase in the relative drop in bearing capacity at selected points of measurement,
4. The reduction of the bending stiffness EJy was determined at an individual corrosion rate.

Stainless steel, thanks to its chemical composition, protects structural elements from atmospheric corrosion and from corrosion caused by 5% sulfuric acid for 720 hours, but does not protect against acids with higher concentrations.

References

[1] G. Łagoda: Wiadukty nad autostradami, Wybrane zagadnienia kształtowania konstrukcyjnego i estetycznego. OW PW Warszawa (2001).

[2] P.Grzesiak, R.Motława, Korozja stali w kwasie siarkowym. Instytut Ochrony roślin, PIB, Poznań (2008).

[3] P.R. Roberge: Corrosion Engineering. Principle and Practice. McGraw-Hill, USA (2008).

[4] PN-EN 10155:1997 + Apl: (2003): Skład chemiczny i wybrane własności mechaniczne stali konstrukcyjnych trudno rdzewiejących.

[5] PN-EN 10088-1. Stale odporne na korozję – część 1.Gatunki stali odpornych na korozję.

[6] K. Wasilewska: Wpływ korozji na nośność wybranych elementów stosowanych w sieciach komunalnych, Rozprawa doktorska. PW, Warszawa (2014).

[7] S.D. Cramer, B.S. Covino (ed.), Corrosion: Environments and Industries, Handbook, Volume 13, ASM International (2006).

[8] H. Szydłowski: Teoria pomiarów, PWN, Warszawa (1981).

[9] A. Glinicka, C. Ajdukiewicz, S. Imiełowski, Effects of uniformly distributed side corrosion on thin-walled open cross-section steel columns. Roads and Bridges, 159 (2016), pp. 257 – 270. https://doi.org/10.1016/j.proeng.2015.07.087

[10] A. Glinicka, M. Kruk, Analiza eksperymentalna wyboczenia niesprężystego prętów stalowych poddanych korozji w komorze mgły solnej. Drogi i Mosty 3 (2010), pp. 5- 27.

[11] A. Glinicka, M. Maciąg, Reduction of the bearing capacity of a thin- walled steel beam-column as a result of uniform corrosion, MATEC Web of Conferences 196, 01046 (2018), XXVII R-S-P seminar 2018, Theoretical Foundation of Civil Engineering. https://doi.org/10.1051/matecconf/201819601046

[12] K. Wasilewska, A. Glinicka, Próba ściskania rur ze stali nierdzewnej poddanych działaniu środowisk agresywnych. Logistyka 3 (2012), pp. 2375 – 2381.

[13] Instrukcja obsługi programu BlueHill v.2.6, Instron Corporation, Copyright, (2004).

Experimental Mechanics of Solids
Materials Research Proceedings **12** (2019) 37-44

Materials Research Forum LLC
https://doi.org/10.21741/9781644900215-5

Investigation on Microstructure and Mechanical Properties of AA 2017A FSW Joints

Dorota Kocanda[1, a], Janusz Mierzynski[2,b], Janusz Torzewski[3,c*]

[1,2,3] Military University of Technology, 2 Witolda Urbanowicza Street, 00-908 Warsaw, Poland

[a]dorota.kocanda@wat.edu.pl, [b]janusz.mierzynski@wat.edu.pl, [c] janusz.torzewski@wat.edu.pl

Keywords: FSW, FSW Joints, 2017A Aluminum Alloy, Mechanical Properties

Abstract. The present paper aims to analyse the microstructure, mechanical properties and fatigue behavior of FSW butt joints. The influence of welding parameters on the quality of butt welds was appreciated in the macro and micro scales on the basis of observations made by means of a confocal microscope. Mechanical properties and HCF fatigue behavior of both the AA 2017A parent material and FSW butt welded specimens were examined experimentally. Tensile properties such a yield strength, ultimate tensile strength and elongation were derived as well. The results of tensile examinations of FSW butt joints were presented in the form of stress-strain curves. On the basis of microstructure analysis and tensile strength tests carried out for FSW butt joints manufactured at few sets of process parameters the most favorable parameters of FSW process were selected. Then, HCF tests were carried out under selected FSW process parameters for the samples either made of wrought material (parent material) or its butt joints. The results of the HCF tests were displayed such as the comparative plots drawn for the AA 2017A alloy specimen and its FSW butt joints.

Introduction

Aluminum alloys find application in various industries, namely aviation, aerospace, automotive, maritime and rail industries [1,2]. These alloys belong to structural materials that meet an increased requirements of modern industry and an environmental protection. Low density of aluminum alloys (2.7 Mg/ m^3), one third of that of steel, allows to significantly reduce the total weight of a structure, and therefore reduce its production costs, fuel consumption and exhaust gas production as well as CO_2 emissions during operation. The accessibility of aluminum in nature, its price, plasticity and high machining of the structural elements make aluminum alloys competitive with modern composites. These issues have become the criteria in the selection of material as well as the development of efficient technologies for joining structural elements in many industries.

Among the available aluminum alloys (1XXX to 7XXX series), the 2XXX series alloys exhibit the highest static and fatigue strength due to the high copper content (2.9-3.5%) in the chemical composition of alloys, which increases the mechanical properties and the precipitate mechanism of material strengthening. It also increases hardness, formability and material strength at elevated temperatures. Copper reduces corrosion resistance, ductility and weldability [3]. Additionally, a critical review of the advancements in joining methods of aluminium alloys has been performed. Aluminum alloys of the 2XXX series are materials that are difficult to weld by conventional methods due to their high thermal conductivity which is about four times that of low-carbon steel. However, the high thermal conductivity of aluminium alloy helps to solidify the molten weld pool of aluminium and, consequently, facilitates out-of-position welding. The strength of welded joints is at the level of 50-90% of the strength of the parent material, due to the brittleness and porosity of the weld and the tendency to microcracks development [4].

Experimental Mechanics of Solids   Materials Research Forum LLC
Materials Research Proceedings **12** (2019) 37-44   https://doi.org/10.21741/9781644900215-5

In order to increase the strength of aluminum joints in 1991, the British Welding Institute (TWI) developed and patented a competitive and ecological technology of frictional joining of aluminum components with material mixing within the weld, known as FSW (Friction Stir Welding) [5] FSW is a solid-state joining process with many advantages, such as robust mechanical and fatigue properties and a narrow heat-affected zone compared with conventional fusion welding [6-7]. The influence of FSW welding parameters on the microstructure changes in the combination of 3 mm thick flat rectangular samples from the AA2024-T8 alloy and the greatness of the generated deformations as well as mechanism of quasi-static cracking of the alloy AA2024-T8 under the influence of stretching was investigated in [8] as well. Effect of rotation speed (400, 600, 800, 1000, 1250 and 1500 rpm) at a constant welding speed (50 mm/min) on microstructure, hardness distributions and tensile properties of the AA2024-T3 alloy plates of 3mm thickness and their FSW joints were analyzed in [9]. It was turned out that increasing rotation speed resulted in finer and more homogenous distribution of second phase particles in the stir zone. A study of fatigue crack growth behavior of FSW welded butt joints of AA2024-T3 aluminium commonly used in riveted aeronautic fuselage structures was presented in [10]. There, is a need to join components with different thicknesses (3.8 mm and 4 mm) using FSW. Crack growth tests on these joints are not standard. Fatigue crack growth rate was analyzed in different zones of the FSW welded joint and in parent material. Paper [11] review the latest developments in the FE numerical analysis of FSW processes, microstructures of FSW joints and properties of FSW welded structures made of different aluminium alloys. Some important numerical issues such as materials flow modeling under heat inputs, meshing procedure, temperature and residual stress distributions in the particular zones of FSW joints and failure criteria are discussed. Authors of the paper [12] paid attention only on the shear zone of a FSW weld considering a mechanism acted there. It can be observed that material in this zone is subjected to different thermo-mechanical cycles. They ask the question, whether this mechanism of shear zone formation results from a temperature increase with higher rotation speed and/or material held for an increased time at temperature, is still not understood. The study does give insight into the often conflicting results published regarding the microstructural evolution in a FSW of AA 2XXX- age hardened alloys.

Experimental procedure
Rectangular plates of dimensions 500x100 mm were cut out from 5 mm thick AA 2017A-T4 heat treated aluminium alloy sheet parallel to its rolling direction. Chemical composition of the aluminium alloy of wchich the specimen was made is presented in Table 1.

Table 1. Chemical composition

wt. %	Si	Fe	Cu	Mn	Mg	Cr	Zn	Ti
2017A	0.2-0.8	0.6-0.8	3.5-4.5	0.4-1	0.4-0.8	0.1-0.12	0.23-0.27	0.17-0.23

The plates have been butt welded parallel to their rolling directions by applying friction stir welding technique and the ESAB Legio 4UT professional device (Fig. 1a). The FSW tool geometry consisted of a threaded conical pin (diameters from 6.5 to 8.7 mm and height 4.8 mm) and a spiral shoulder with a diameter of 19 mm (Fig. 1b). The tool has been tilted at an angle of 2 degrees. The joining process was carried out in the position of control mode.

Before joining process, the plates were subjected to grinding to remove alumina oxides from their surfaces by means of the grinding paper of different granularities. Then, the plates were cleaned.

Experimental Mechanics of Solids
Materials Research Proceedings **12** (2019) 37-44

Materials Research Forum LLC
https://doi.org/10.21741/9781644900215-5

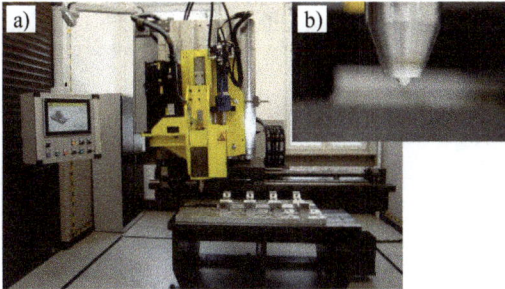

Fig. 1. The ESAB Legio 4UT device (a) and FSW tool geometry (b)

Four different combinations of joining parameters were used in the studies. The welding parameters are provided in Table 2.

Table 2. Welding parameters used in the experiments

Material	Rotation speed [rpm]	– Welding speed [mm/min]		
AA 2017A-T4	500 - 200	500 - 400	350 - 200	350 - 400

In order to derive mechanical and fatigue properties the specimens of geometry prepared according to ISO 6892-1:2009 were cut out from the welded plates perpendicular to the welding line. The way of taking specimens from the sheet and their sizes are shown in Fig. 2. The test specimens were not specially prepared except for the removal of the material that had flowed out at the joint. The surface from the front of the joint was remained after the joining process (Fig.7).

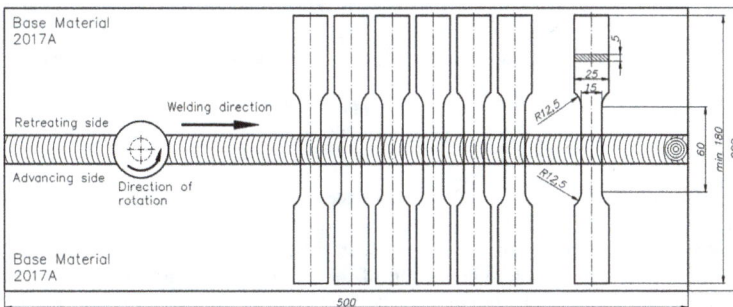

Fig. 2. Arrangement of AA 2017A sheets at the time of making FSW joints and dimensions of specimens in mm used for strength tests

Metallographic analyses were conducted on the specimen's cross-sections perpendicular to the welding direction. The cross-sections of specimens were polished using 3 and 1 μm diamond solution and then etched with reagent consist of hydrofluoric acid, nitric acid and water for about 10 s. Observations of the base material and the welds were carried out using the Olympus LEXT OLS 4100 laser scanning confocal microscope. The Vickers hardness measurements were carried out in the cross-section of the joint in the centre of the specimen thickness using the Struers Dura Scan 70 device. The load of 0.98 N was applied for 10 s at a measuring point spacing of 0.5 mm.

Experimental Mechanics of Solids   Materials Research Forum LLC
Materials Research Proceedings **12** (2019) 37-44   https://doi.org/10.21741/9781644900215-5

Tensile and fatigue tests were carried out on an INSTRON 8802 universal testing machine with WaveMatrix computer software. Fatigue life tests were carried out under constant load amplitude control, at R = 0.1 and 12 Hz frequency. Fatigue failure was used as a criterion for the end of the test.

Experimental results
The FSW joints of the AA 2017A-T4 plates under examinations, obtained at four different combinations of process parameters were subjected to microstructural analysis.

Fig. 3 presents a macroscopic image of the FSW weld made at 500 rpm of rotation speed and 200 mm/min of welding speed. It can be observed there that the weld arose as an effect of the simultaneous actions of complex thermomechanical processes. As a result, the microstructure was modified in particular zones of joints. In their central part of the weld is located the nugget zone (NZ) where the mechanical-thermal interactions were the greatest and the phenomenon of dynamic recrystallization occurred. As a consequence, there was found a fine-grained microstructure with even-axial grains, the size of which ranged from 4 to 6 μm (Fig. 4a). The shape of this zone and its microstructure strongly depended on the parameters of the joining process used. Three other characteristic zones of the weld, symmetrically located in relation to the nugget zone (NZ), can be clearly distinguished in the cross-section of the FSW joint. There are thermo-mechanical affected zone (TMAZ), heat affected zone (HAZ) and base material (BM).

Fig. 3. Macrograph of the FSW joint cross-section referred to the welding parameters: n=500 rpm and v=200 mm/min

In Fig. 4a–d magnified images of the mentioned zones, namely NZ, TMAZ, HAZ and to the base materials (BM) of AA 2017A alloy are showed. In the adjacent to the nugget zone a narrow thermo-mechanical affected zone was created. In this area the material has been subjected to both a strong mechanical deforming action of the microstructure grains and a thermal action changing the mechanical properties of the material. However, the thermal impact was not as high as in the case of the NZ so no recrystallization process was observed.

The image of the transition zone from NZ to TMAZ is shown in the picture in Fig. 4b. At a greater distance from the NZ, the mechanical impact on the material was smaller. However, a strong influence of temperature was still observed. This was reflected in the image of the microstructure shown in Fig. 4c and called as heat affected zone HAZ. In this zone no visible signs of plastic deformation were observed, nonetheless, the grain size of the microstructure is clearly larger than in the base material (BM). The microstructure of the base material presented in Fig. 4d was characterized by a clear texture resulting from the rolling of sheets in the production process.

Experimental Mechanics of Solids
Materials Research Proceedings **12** (2019) 37-44

Materials Research Forum LLC
https://doi.org/10.21741/9781644900215-5

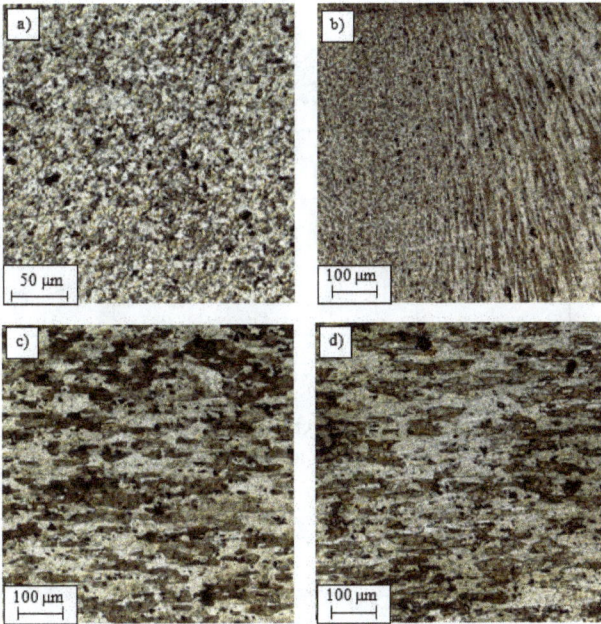

Fig. 4. Microstructures observed in the characteristic zones of the FSW joint: nugget zone (a), thermo-mechanical affected zone (b), heat affected zone (c) and base material (d).

The results of Vickers microhardness tests conducted along the cross-section of joints for three different sets of FSW process parameters are shown in Fig. 5. The measurements were carried out in a line lying in the middle of the sample thickness. The microhardness profile is characteristic for aluminium alloys showing precipitated hardening. One can notice in the joint area a significant decrease in the microhardness course below the microhardness of the base material of value equal to 134 $HV_{0.1}$ (horizontal dashed line in the graph). This drop is attributed to the dissolution of the alloy precipitated hardening as a result of the strong thermal pulse generated by the FSW process. The distribution of microhardness in the zones of the joints made at n = 500 rpm and welding speeds of 200 and 400 mm/min respectively, resembles the letter "W", slightly offset from the central axis of the joint towards the retreating side of the weld. The minimum µHV values are between 116-118 $HV_{0.1}$ on the material retreating side and around 120-121 $HV_{0.1}$ on the advancing side. These ones are located in HAZ zones of the joint.

It is worth emphasizing that the microhardness in the nugget zone (NZ) is clearly higher than the minimum values, although it does not reach the µHV values typical for the base material. In the case of a joint made at n = 350 rpm and v = 400 mm/min, the distribution of microhardness is different. The minimum hardness was obtained in the NZ zone and it fluctuated in the range of 98-115 $HV_{0.1}$. Such a course of microhardness resulted from incorrect parameters of the FSW process for this alloy, which was confirmed by strength tests.

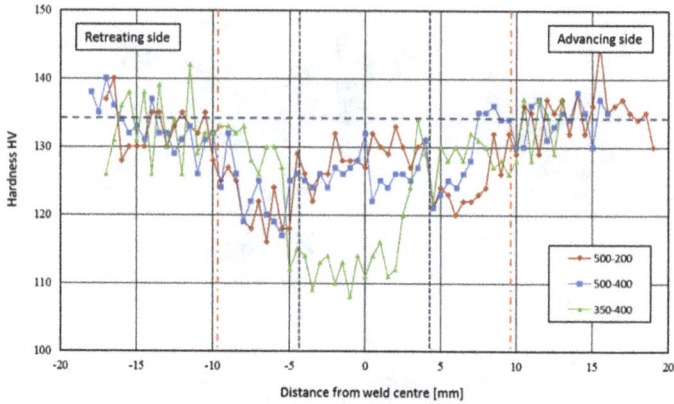

Fig. 5. Microhardness profiles across the FSW welds obtained from the specimens' cross-sections at various welding parameters

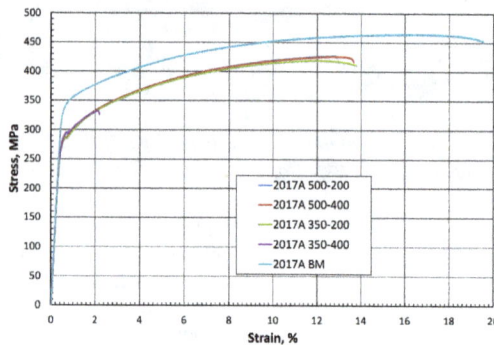

Fig. 6. Comparative plots of stress against strain derived from the tensile tests curried out for the AA 2017A-T4 alloy (base material) and its FSW butt joints at different values of process parameters

The monotonic tensile tests of the base material and butt joint specimens were carried out in order to identify the influence of joining parameters on the basic mechanical properties of standardized samples made of AA 2017A. The analysis of the test results made it possible to determine the basic mechanical parameters, i.e. offset yield stress ($R_{p0.2}$), ultimate tensile stress (R_m) and relative elongation at break (A). The results of strength tests in the form of stress-strain curves drawn both for the base material and for FSW welded samples manufactured at different process parameters are presented in Fig. 6. It can be noted that FSW butt welded specimens show a significant decrease in elongation at break (A) in relation to the parent material. This decrease of about 5.5 % has been recorded for three sets of process parameters. In the case of the specimen welded at the parameters n = 350 rpm and v = 400 mm/min, the mentioned above drop was even greater and the maximum deformation was only 2%. The results of static tensile test conducted for the base material and FSW joints manufactured at four sets of process parameters are presented in Table 2. Additionally, Table 2 shows the relative strength coefficient as the ratio of tensile strength of FSW joint (R_{mFSW}) to the strength of the base material (R_{mB}). This parameter clearly indicates that for a properly executed joint using the FSW method, a slight reduction of strength properties of amount 6-7 % is recorded. However, if the parameters have been wrongly selected, the decrease of strength properties was significant and amounted to about 27%.

Table 2. Mechanical properties of AA 2017A

	$R_{p0,2}$ (MPa)	R_m (MPa)	A (%)	R_{mFSW}/ R_{mB}
2017A Base Material	330	450	19.0	
2017A 500 - 200	285	425	13.0	0,94
2017A 500 - 400	290	425	13.5	0,94
2017A 350 - 200	285	420	13.5	0,93
2017A 350 - 400	290	330	2.0	0,73

Similar results were obtained for stresses corresponding to the offset yield stress ($R_{p0.2}$). The stresses in the welded samples were about 10 % lower than in the base material for all sets of joining parameters (see Table 2).

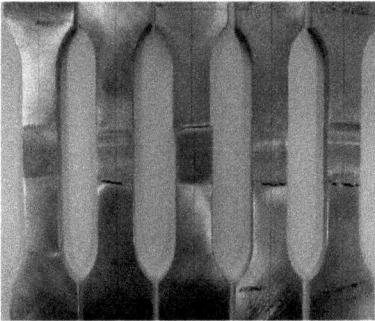

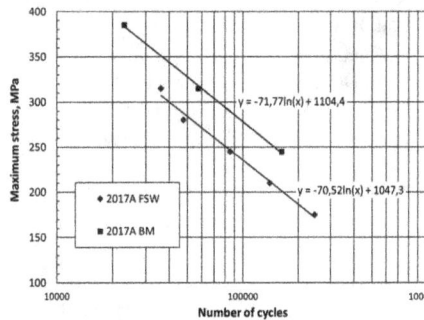

Fig. 7. General view of broken FSW welded samples after HCF tests.

Fig. 8. HCF test results derived for wrought AA 2017A alloy and for FSW joints.

Preliminary HCF fatigue tests have been conducted for the base material as well as for the FSW welded specimens at n = 500 rpm and welding rate of v = 400 mm/min. Fatigue tests were carried out in the limited fatigue life region at the stress ratio R = 0.1. The images of broken samples after HCF tests were showed in Fig. 7. It can be observed that the cracking process of the welded samples proceeded mainly in the weld and along the boundary between the hardened area and the parent material. This clearly proves the good quality of the FSW joint and the dominant of the structural notch effect over the technological notch. Comparative SN characteristics of HCF life seemed in Fig. 8 correspond to the samples made of base material (BM) and to FSW joints (FSW) and tested in the range of about $N_f = 2 \cdot 10^4 \div 3 \cdot 10^5$ number of cycles. The mentioned diagram indicates lower fatigue life of welded samples than this one for the parent material.

Summary

The study proved that FSW technology allows to obtain good quality joints featured by good strength properties for the 2017A aluminium alloy in a wide range of process parameters. The microhardness profiles measured along the specimens cross-sections showed characteristic shape "W" for aluminium alloys hardened by precipitates. Significant decrease in microhardness parameters was observed in the case of improperly selected joint parameters, as well. FSW joints manufactured at parameters n = 500 rpm and v = 200 mm/min were characterized by the best mechanical properties and microstructure. The study revealed that the ultimate tensile strength of FSW welded specimens reached 94% values of the strength corresponded to the base material. On the other hand, joints produced at parameters n = 350 rpm and v = 400 mm/min showed poor mechanical and fatigue properties due to numerous defects resulting from improperly selected parameters of the joining process.

References

[1] J. Hirsch, Recent development in aluminium for automotive applications. Trans. Nonferrous Metals Soc. China 2014, 24, 1995–2002. https://doi.org/10.1016/s1003-6326(14)63305-7

[2] P. Rambabu, N.E. Prasad, V.V. Kutumbarao, R.J.H. Wanhill, Aluminium alloys for aerospace applications. In Aerospace Materials and Material Technologies; Prasad, N.E., Wanhill, R.J.H., Eds.; Springer: Singapore, 2017; pp. 29–52. https://doi.org/10.1007/978-981-10-2134-3_2

[3] R. Rajan, P. Kah, B. Mvola et al., Trends in aluminium alloy development and their joining method, Rev. Adv. Mater. Sci., 44, (2016), 383-397.

[4] G. Cam, G. Ipekoglu, Recent development in joining of aluminium alloys, Int. J. Adv. Manuf Technol, December, 91, (2017), 1851-1816.

[5] W.M. Thomas, J.C. Nicholas, M.G. Needham, T. Smith, C.J Dawes,. Friction Stir Butt Welding. International Patent Application No. PCT/GB92/0220, December 1991

[6] R.S. Mishra, Z.Y. Ma, Friction stir welding and processing, Mater. Sci. Eng. R Rep. 50 (1–2) (2005) 1–78.

[7] H.J. Liu, H. Fujii, M. Maeda, K. Nogi, Tensile properties and fracture locations of friction-stir-welded joints of 2017-T351 aluminum alloy. J. Mater. Process. Technol. 2003, 142, 692–696. https://doi.org/10.1016/s0924-0136(03)00806-9

[8] P. L. Threadgill, A. J. Leonard, H. R. Shercliff, P. J. Withers; Friction stir welding of aluminium alloys, Int Materials Reviews, vol. 54(2009), 49-93. https://doi.org/10.1179/174328009x411136

[9] S. A. Khodir, T. Shibayanagi, M. Naka, Microstructure and mechanical properties of friction stir welded AA2024-T3 aluminium alloy, Mater. Transactions, 47,1, (2006), 185-193. https://doi.org/10.2320/matertrans.47.185

[10] P.M.G.P. Moreira, P.M.S.T de Castro, Fatigue crack growth on FSW AA2024-T3 aluminim joints, Key Engineering Materials, 498, (2012), 126-138. https://doi.org/10.4028/www.scientific.net/kem.498.126

[11] Xiaocong He, Fengshou Gu, A. Ball, a review of numerical analysis of friction stir welding, Progress in Materials Science, 65, (2014), 1-66. https://doi.org/10.1016/j.pmatsci.2014.03.003

[12] J. Schneider, R. Stromberg, P. Schilling et al., Processing effects on the friction stir weld stir zone, Welding Journal, 01, (2013), 1-9.

Experimental Mechanics of Solids
Materials Research Proceedings **12** (2019) 45-51

Materials Research Forum LLC
https://doi.org/10.21741/9781644900215-6

Analysis of the Influence of Mechanical Couplings in Laminate Beams on the Adherence of the Assumed Boundary Conditions in the DCB Test Configuration

Jakub Paśnik[1, a *], Sylwester Samborski[1,b], Jakub Rzeczkowski[1,c] and Katarzyna Słomka[1, d]

[1]Lublin University of Technology, Department of Applied Mechanics, 20-618 Lublin, Nadbystrzycka 36 St., Poland

[a] jakub.pasnik@pollub.edu.pl, [b]s.samborski@pollub.pl, [c]kubarzeczkowski@op.pl, [d]katarzyna.slomka@onet.com.pl

Keywords: Composite Laminate, Finite Element, Delamination, Double – Cantilever Beam, VCCT

Abstract. This paper shows numerical analyses of delamination in coupled laminates using finite element method. With increasingly wide spreading use of laminate composites a research development of those materials goes on. It is a common knowledge that the main form of damage in composites is delamination that is a loss of cohesion between neighboring layers. The main aim of the conducted research is to obtain the strain energy release rate (SERR) distributions along initial delamination front and verifying compliance of experimental and numerical analyses. Some analyses were performed using numerical models based on the double cantilever beam (DCB) test configuration, because it allowed to determine the values of SERR in mode I, G_I. The analyses were carried out using the Abaqus/CAE finite element software environment. Models of the specimens to be tested experimentally were elaborated in accordance with the DCB test configuration for which the boundary conditions and the load were specified. To model delamination process in composite beams the virtual crack closure technique (VCCT) was used [1]. In addition, numerical analyses of the boundary conditions and the laminate stacking sequence effect on the SERR distribution were done [1]. What is more, experimental analyses of crack shape were carried out. The results were obtained for two kinds of coupled layups: bending – extension and bending – twisting, as well as for uncoupled specimens and were compared mutually: the coupled and the uncoupled layups. The results of tests showed significant influence of boundary conditions and laminate stacking sequence both on the G_I distribution along delamination front and on the crack front shape. It was noted that couplings have considerable impact on both numerical analyses' results and can induce unwanted deformations of the test specimens during physical experiments. Moreover, crack shapes in numerical models and in experimental specimens were compared.

Introduction

Composite materials are being increasingly used as construction materials. They can often be a substitute of conventional materials, such as steel or aluminum, used for making specific construction elements. What is it that distinguishes these materials from a wide range of materials? These are in particular their specific characteristics. Composites are the materials that contain two components called phases: the matrix phase and the reinforcing phase (in this instance – fibers). The matrix is supposed to connect elements of reinforcing phase and to retain some specified fibers orientation in the whole volume of element. The reinforcing phase can take form of fibers or molecules. Due to combination of two materials of significantly different

Experimental Mechanics of Solids
Materials Research Proceedings **12** (2019) 45-51

Materials Research Forum LLC
https://doi.org/10.21741/9781644900215-6

parameters, a new material has extraordinary parameters which are impossible to obtain in conventional materials. The main advantage of laminate composites, which are considered in this paper, is great toughness with relatively low weight. When one compares use of composite material with use of steel or aluminum, he will notice not only better durability but primarily significant weight reduction of designed element. And these all advantages determine use such materials in range of industrial sectors such as: automotive, aviation, aerospace, civil engineering but also as components of sport equipment [4]. A wide range of application of composites requires precise research on parameters of these materials, mechanisms and terms of damage. This paper is to analyze the most common damage in laminate composites – delamination, which is loss of cohesion between neighboring layers. There were elaborated three modes to study conditions of fracture propagation in construction materials, they are schematically shown in fig. 1.

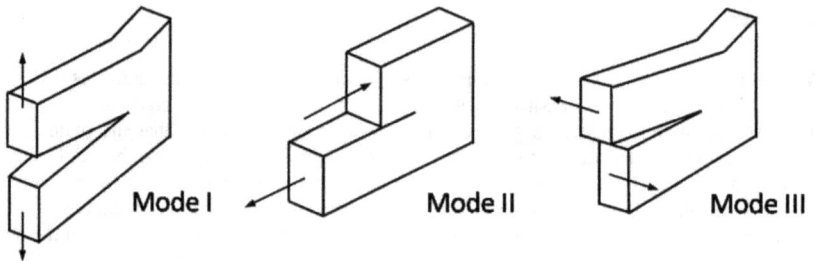

Fig. 1. Fracture modes

Parameters of tested beams
This paper shows numerical and experimental analyses of specimens made out of laminate composite of the commercial name *SEAL Texipreg HS160RM*. Material characterization is shown in Table 1.

The specimens used for tests were prepared precisely. Beams were made out of prepregs in the shape of unidirectional reinforcement. Single carbon—epoxy ply thickness was 0,1 mm, and glass ply thickness equals 16 carbon—epoxy plies thickness. In tested beams initial crack was introduced by implementing PTFE foil between two branches of specimen on the length of 65mm (Fig. 2). In order to make crack front shape visible (in experiment), white paint was injected pressurized beetwen separated branches. Paint was supposed to fill crack and show how far crack front went during test. To observe crack front shape, two branches of beam was separated completely which made it possible to monitor crack shapes.

Experimental Mechanics of Solids
Materials Research Proceedings **12** (2019) 45-51

Materials Research Forum LLC
https://doi.org/10.21741/9781644900215-6

Table 1. Material properties

Basic material constants					
E_1 [MPa]	$E_2=E_3$ [MPa]	$v_{12}=v_{13}$ [-]	v_{23} [-]	$G_{12}=G_{13}$ [MPa]	G_{23} [MPa]
109000	8819	0,342	0,380	4315	3200
Fracture mechanics constants			
G_{Ic} [N/mm]	G_{IIc} [N/mm]	G_{IIIc} [N/mm]	η [-]
0.4	0.8	0.8	1.62

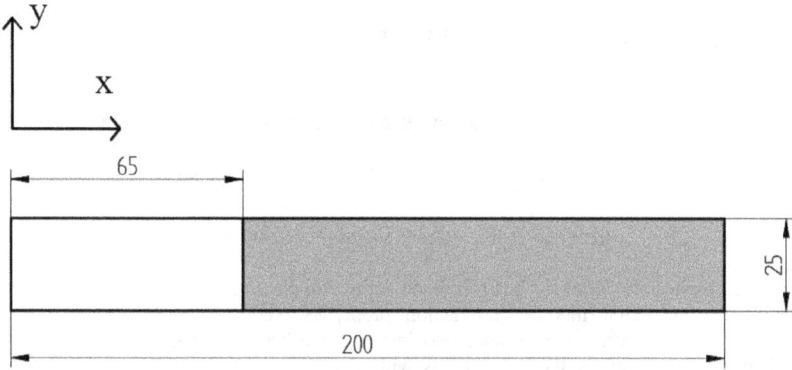

Fig. 2. Initial crack length

Numerical modeling technique
Experimental research results were confronted with results obtain in numerical tests. The numerical analyses have been performed in the Abaqus/CAE finite element software environment. Utilization on such software made it possible to make numerical model of the beam and conduct simulations reflecting actual experimental tests. Geometric model of laminate was prepared as a shell element to which composite layups was implemented in *Property module*. Specimen was tested in mode I with boundary condition reflecting double cantilever beam [1]. Boundary condition and loading of the beam was presented schematically on fig. 3.

Experimental Mechanics of Solids
Materials Research Proceedings **12** (2019) 45-51

Materials Research Forum LLC
https://doi.org/10.21741/9781644900215-6

Edge 1-2-3 of lower branch was not able to move – all degrees of freedom were disabled. To edge 4-5-6 vertical displacement δ was attached along z axis. The other degrees of freedom were also disabled with exception of rotation along y axis.

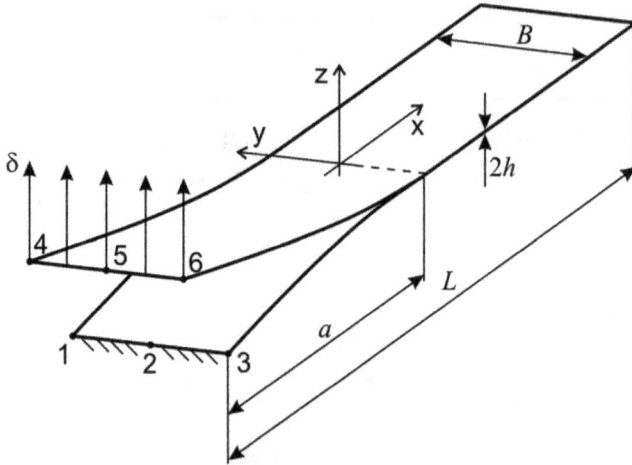

Fig. 3. Boundary conditions and loading [1]

With the translation vector defined as u_i and rotation as θ_i, boundary conditions can be submitted as:

$$u_{ix} = u_{iy} = u_{iz} = \theta_{ix} = \theta_{iz} = 0 - \text{for } 1\text{-}2\text{-}3 \text{ edge,}$$

$$u_{ix} = u_{iy} = \theta_{ix} = \theta_{iz} = 0, \ u_{iy} = \delta - \text{for } 4\text{-}5\text{-}6 \text{ edge.}$$

Numerical model of the beam is made out of two parts connected with surface-to-surface contact in Interaction module. Interaction properties were shown in table 1. To avoid convergence problems, small distance was implemented between parts with *Clearance* option. Each instance was meshed with elements with size of approx. 2 mm. However, area where delamination was supposed to occur and near the beam's edges was divided on more elements with size of 0,5 mm. Such mesh density was relevant impact of numerical results. Elements types used in simulation were S4 near delamination front and S4R where propagation was not expected.
For delamination analysis Virtual crack closure technique (VCCT) was used. As fracture criterion the Reeder Law has been chosen [3]:

$$G_{eq-c} = G_{Ic} + (G_{II} - G_I)\left(\frac{G_{II} + G_{III}}{G_T}\right)^{\eta} + (G_{IIIc} - G_{IIc})\left(\frac{G_{III}}{G_{II} + G_{III}}\right)\left(\frac{G_{II} + G_{III}}{G_T}\right)^{\eta} \quad (1)$$

where G_{eq-c} stood for equivalent critical strain energy release rate (SERR)

Experimental Mechanics of Solids   Materials Research Forum LLC
Materials Research Proceedings **12** (2019) 45-51   https://doi.org/10.21741/9781644900215-6

$$G_T = G_I + G_{II} + G_{III}$$

$$(2)$$

When G_T exceeds G_{eq-c} delamination occurs.

Results and discussion

In this paragraph results of numerical and experimental tests were presented.

The fig. 4 below presents crack front shapes in laminate composites. Crack front shapes obtainin numerical analyses and experimental testswas compared. As can be seen, both numerical and experimental analyses' results correspond with each other in specimens no. 1 and 2. Fronts have similar shapes for these tests. Although, specimens 3 and 4 tests give dissimilar results. Numerical analysis of beam no. 3 shows reverse curvature than experimental work does. In experimental test of specimen 4 on the other hand, there were some manufacturing error, which cause very unusual front shape. However, it should be noted that during tests, some unwanted effect occurs. It is when two branches did not separate completely during test and are still linked with partly separated, decussate fibers. And it is called bridging phenomenon.

In addition to that, SERR distributions along delamination front were obtained in FEM analyses.

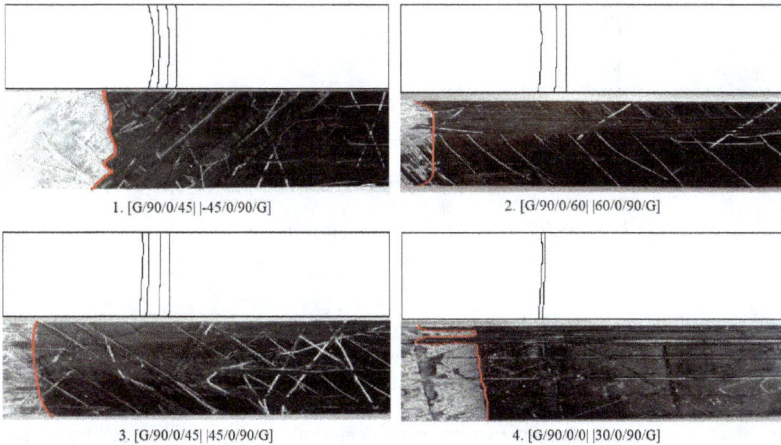

1. [G/90/0/45| |-45/0/90/G]

2. [G/90/0/60| |60/0/90/G]

3. [G/90/0/45| |45/0/90/G]

4. [G/90/0/0| |30/0/90/G]

Fig. 4. Crack front shapes comparison

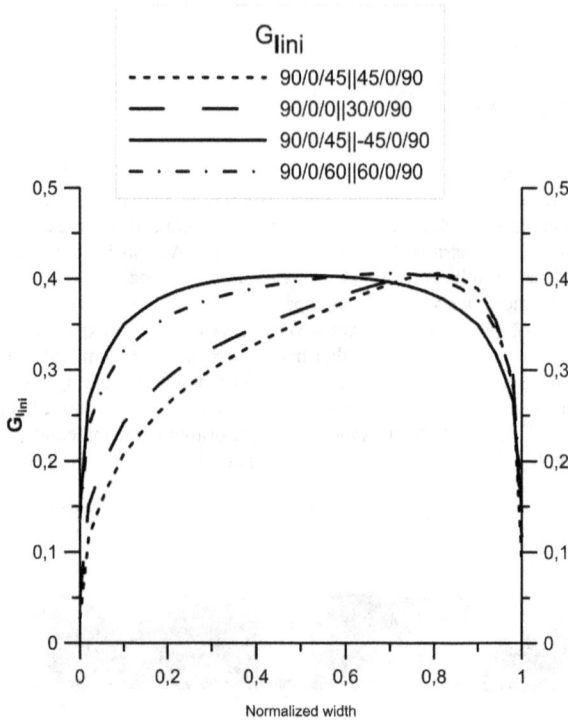

Fig.5 SERR distribution along delamination front

Summary

In this a paper comparison of numerical modeling results with experimental tests results in the field of crack front shape in laminate composites was presented. The results verified accuracy of using VCCT for modeling delamination with ABAQUS/CAE finite element environment. In addition to that, some problems which can appear in experimental tests were observed, such like bridging effect or some manufacturing errors. Numerical results in the form of crack front shapes and SERR distributions were verified in experimental test in different stacking sequences. Further research will aim to optimize numerical model and reduce impact of unwanted effects in experimental tests.

Acknowledgements

This paper was financially supported by the Ministerial Research Project No. DEC-2016/21/B/ST8/03160 financed by the Polish National Science Centre.

References

[1] S. Samborski, Numerical analysis of the DCB test configuration applicability to mechanically coupled Fiber Reinforced Laminated Composite beams, Composite Structures 152 (2016) 477 – 487. https://doi.org/10.1016/j.compstruct.2016.05.060

Experimental Mechanics of Solids
Materials Research Proceedings **12** (2019) 45-51

Materials Research Forum LLC
https://doi.org/10.21741/9781644900215-6

[2] S. Samborski, Analysis of the end-notched flexure test configuration applicability for mechanically coupled fiber reinforced composite laminates, Composite Structures 163 (2017) 342 – 349. https://doi.org/10.1016/j.compstruct.2016.12.051

[3] S. Samborski, J. Rzeczkowski, J. Paśnik, Issues of Direct Application of Fracture Toughness Determination Procedures to Coupled Composite Laminates, IOP Conference Series: Materials Science and Engineering, Volume 416 (2018) 012056. https://doi.org/10.1088/1757-899x/416/1/012056

[4] J. Paśnik, S. Samborski, J. Rzeczkowski, Application of the CZM Technique to Delamination Analysis of Coupled Laminate Beams, IOP Conference Series: Materials Science and Engineering, Volume 416 (2018) 012075. https://doi.org/10.1088/1757-899x/416/1/012075

[5] C.B.York, Unified Approach to the Characterization of Coupled Composite Laminates: Benchmark Configurations and Special Cases, Journal of Aerospace Engineering 23(4) (2010). https://doi.org/10.1061/(asce)as.1943-5525.0000036

[6] M.F.S.F. De Moura, R.D.S.G. Campilho, J.P.M. Gonçalves, Crack Equivalent Concept Applied to the Fracture Characterization of Bonded Joints under Pure Mode I Loading. Composites Science and Technology 68 (2009) 2224. https://doi.org/10.1016/j.compscitech.2008.04.003

[7] V. Burlayenko, T. Sadowski, FE modeling of delamination growth in interlaminar fracture specimens, Budownictwo i Architektura 2 (2008) 95 – 109.

[8] J. German, Podstawy mechaniki kompozytów włóknistych, Wyd. PK, Kraków, 2001

[9] ABAQUS Online Documentation, Version 6.14,© DassaultSystèmes, 2014.

Experimental Mechanics of Solids
Materials Research Proceedings **12** (2019) 52-58

Materials Research Forum LLC
https://doi.org/10.21741/9781644900215-7

Comparison of Different Methods for Determination of Delamination Initiation Point in the DCB Test on Coupled CFRP Composite Laminates

Jakub Rzeczkowski[1, a], Sylwester Samborski[1,b*]

[1] Lublin University of Technology, Department of Applied Mechanics, Nadbystrzycka 36 St, Lublin 20-618, Poland

[a]kubarzeczkowski@op.pl, [b]s.samborski@pollub.pl

Keywords: Damage, Delamination, Double Cantilever Beam, Acoustic Emission, Mechanical Coupling

Abstract. This article presents an experimental determination of the mode I critical strain energy release rate (c-SERR) for different initiation definitions. The multidirectional (MD) and mechanically coupled laminates were subjected to the double cantilever beam (DCB) test. The NL, P_{max} and *5%* criteria were taken into account in calculation of the fracture toughness (G_{IC}). An acoustic emission (AE) technique was used as additional initiation criterion. On the basis of the performed test it was found that the critical strain energy release rate depends on the adopted definition as well as on the specimen interface and mechanical couplings.

Introduction

The fiber reinforced plastic (FRP) laminates are widely used in many industry sectors. The main advantages of polymeric composites are high strength and corrosion resistance. Those materials are used in the manufacturing of the contemporary load carrying structures in airplanes, cars and marine industry [1]. During the operation, on the construction parts different and often variable loads act, as well as various environmental factors must be considered which can cause damage. The most common type of failure in the FRP laminates is called delamination which is the major weaknesses of advanced composite structures. Knowledge of a laminated composite materials resistance to interlaminar fracture is useful for product development and material section. The fracture toughness (G_{IC}) in the form of the critical strain energy release rate (c-SERR) values can be determined for the mode I fracture in the experiment on Double Cantilever Beam (DCB). Such tests are standardized by the ASTM organization (see the ASTM D5528 Standard [2]), where different calculation methods are described. In the available literature, there are some papers addressing to delamination resistance problem for the FRP laminates taking into account the influence of fiber misorientation at delamination interface [3-4], as well as the effect of the mechanical coupling [5-12] on the critical strain energy release rate (c-SERR). One of the problems with precise determination of the mode I c-SERR may be fiber bridging phenomena. This mechanism results from growing the delamination between plies of dissimilar orientation in multiply laminated composite structures. Moreover, in the nonunidirectional DCB specimens a pure mode I fracture may not be achieved as a result of the mechanical couplings [5,8,9]. Another issue in computation of the G_{IC} is properly determination of the initiation point. In the experiments led by the authors of the current paper, the difficulties with precise determination of delamination onset point where experienced. The knowledge of the actual value of the load corresponding to delamination onset is however of particular importance while computing the mode I fracture toughness in accordance with the respective ASTM Standard. Any uncertainty in experimental reading of the fracture onset load can lead to generating the inaccurate results in

Experimental Mechanics of Solids   Materials Research Forum LLC
Materials Research Proceedings **12** (2019) 52-58   https://doi.org/10.21741/9781644900215-7

the G_{IC} calculations. Therefore, four different definitions of initiations were used by the authors to calculate the critical strain energy release rate with the experimental results: the peak force (P_{max}), the deviation from the linearity point force (NL), the 5% compliance increase force, as well as the acoustic emission (AE) indicated force. To obtained mode I c-SERR are the results of two calculation methods: the Modified Beam Theory (MBT) and the Compliance Calibration Method (CCM). The experimental results for the four different approaches of determination of the peak load are discussed in this paper.

Test specimens
Experimental DCB tests was performed on rectangular uniform thickness carbon epoxy composite laminates. The geometrical dimensions of the manufactured specimens was width b=25mm, thickness h=5mm and total length equal 175mm. The composite samples contained nonadhesive insert on the mid plane that serves as a delamination initiator. The scheme and geometrical dimensions of the DCB specimens was presented in Fig. 1.

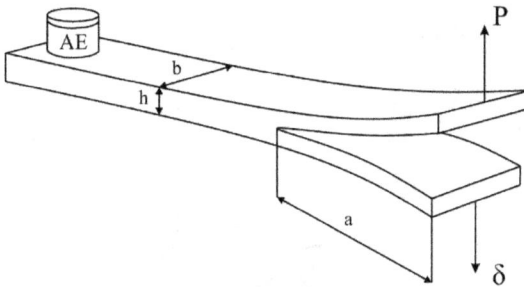

Fig. 1. DCB deformed specimen with AE sensor.

During the tests the multidirectional laminates with delamination interfaces [30°/30°] and [45°/45°] were used. Additionally, bending-twisting (BT) coupled laminate with the respective ply sequence: [45°/0°/45°/45°/0°/-45°/0°/-45°/-45°/-45°/-45°/0°/-45°/45°/0°/0°/45°/45°] and bending-extension (BE) coupled laminate [45°/-45°/0°/-45°/0°/ 45°/90°/45°/-45°] were examined.

Experiment
The experiment was performed on the Shimadzu ASG-X tensile testing machine with 5kN load cell according to the ASTM D5528 Standard indications. The samples with sticked piano hinges was mounted between the grips of the loading machine and then aligned and centered. Prior to test, both edges of the specimens was coated white fluid to aid in visual detection of delamination onset. The test was carried out with constant crosshead speed equal 1mm/min which caused quasi-static opening the specimens. During the experiment the applied load P and displacement δ was measured and registered by the trapezium-X software. Delamination onset as well as all propagation values were visually observed and marked on the specimens edges.

Additionally, the acoustic emission (AE) technique was used in order to increase of the accuracy of detect initiation of the peak force values related to the fracture process initiation.

Experimental Mechanics of Solids   Materials Research Forum LLC
Materials Research Proceedings **12** (2019) 52-58   https://doi.org/10.21741/9781644900215-7

The AE signal was registered by the piezoelectric sensor mounted on unloaded end of the specimens. The AE phenomena was used as one criterion in calculation of the mode I critical strain energy release rate.

Initiation criteria

Owing to difficulties in defining the exact instant of crack of crack initiation four definitions for an initiation value of G_{IC} were used. These include mode I critical strain energy release rate values determined using the load and deflection measured at the point of deviation from linearity in the load-displacement curve (NL), at the maximum load values and at the point which the compliance has increased by 5%. The fourth definition is AE phenomena which was mentioned in previous subsection. The NL G_{IC} value is typically the lowest of the four initiation values and is recommended for generating delamination failure criteria in durability and damage tolerance analyses of laminated composite structures. The exemplary load-displacement curve obtained during one DCB experiments with depicted initiation points was presented in Fig. 2.

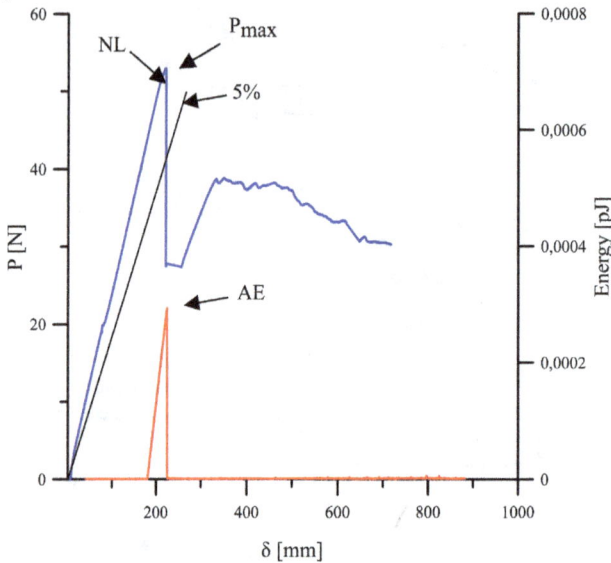

Fig. 2. Exemplary load-displacement curve and energy of AE signal obtained during one of the DCB experiment with determined delamination initiation point definitions.

Calculation methods

To determine the mode I critical strain energy release rate two data reduction methods were used. These consisted of a Compliance Calibration Method CCM and the Modified Beam Theory MBT,

Experimental Mechanics of Solids   Materials Research Forum LLC
Materials Research Proceedings **12** (2019) 52-58   https://doi.org/10.21741/9781644900215-7

The CCM method determines additional parameter n which is the slope of $ln(C)$ versus $ln(a)$ curve. Therefore, the critical SERR is calculating as follows:

$$G_{IC} = \frac{nP\delta}{2ab} \qquad (i)$$

MBT method uses correction parameter Δ which may be determined experimentally by generating a least squares plot of the cube root of compliance $C^{\frac{1}{3}}$ as a function of delamination length a. The load and displacements corresponding to the visually observed delamination onset on the edge and all the propagation values were used to generate this plot. Mode I interlaminar fracture toughness is expressed following equation:

$$G_{IC} = \frac{3P\delta}{2b(a+|\Delta|)} \qquad (ii)$$

Results and discussion

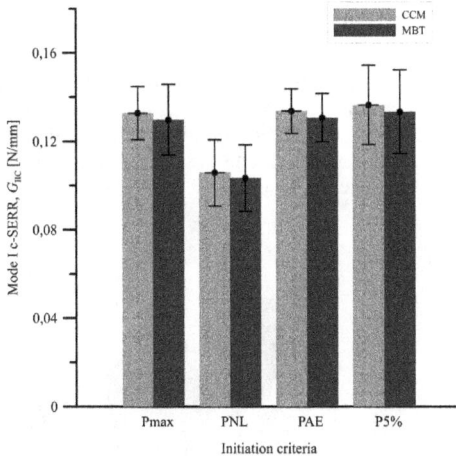

Fig. 3. Mode I c-SERR calculated for different initiation criteria for the specimen with delamination interface 30°/30°

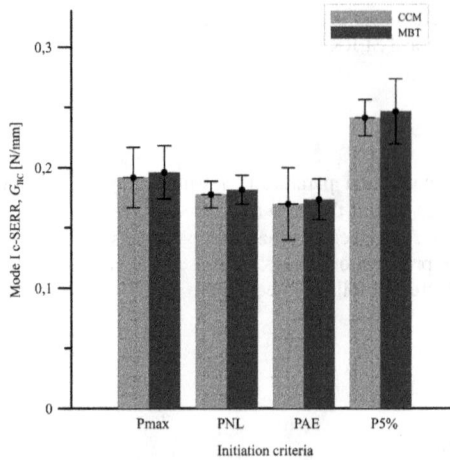

Fig. 4. Mode I c-SERR calculated for different initiation criteria for the specimen with delamination interface 45°/45°

Fig. 3 and Fig. 4 presents the experimentally obtained values of the mode I critical strain energy release rate for specimens with 30°/30° and 45°/45° delamination interface calculated for four different criteria. For both specimen, the G_{IC} reached the greatest values for the 5% initiation definition. Slightly less value of the c-SERR was obtained for the maximum load point criterium which gave similar values like for the AE and the NL initiation points, whereus, for deviation from linearity definitions the values of the G_{IC} were the lowest.

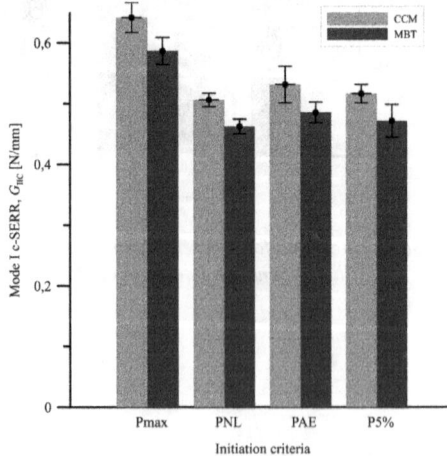

Fig. 5. Mode I c-SERR calculated for different initiation criteria for the specimen exhibiting bending-extension coupling

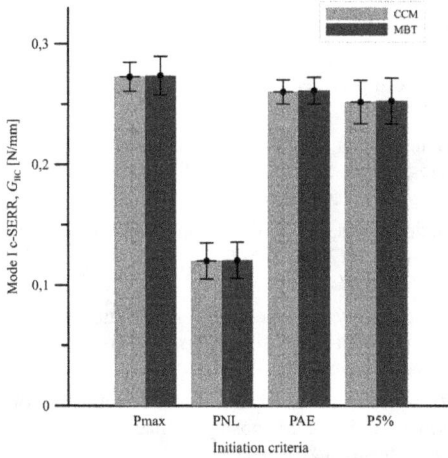

Fig. 6. Mode I c-SERR calculated for different initiation criteria for the specimen exhibiting bending-twisting coupling

Mechanically coupled laminates exhibited different values of the mode I critical strain energy release rate obtained for four different criteria than multidirectional composites. For the bending-extension coupling values of the G_{IC} were on average level about 0.55 N/mm where for bending-twisting coupling this values reached about 0.28 N/mm. The greatest values of the mode I c-SERR was obtained for the P_{max} initiation criterium. The results of calculation of the G_{IC} for the AE and the 5% initiation points were alike and equal about 0.5 N/mm and 0.25N/mm respectively for the BE and the BT specimens. For both samples, the c-SERR exhibited the lowest values for the deviation from linearity delamination initiation point.

For the CCM and the MBT methods the differences between the results of calculations of the mode I c-SERR for all specimens were minimal.

The experimental values of the mode G_{IC} obtained for mechanically coupled laminates for different initiation definitions was depicted in Fig. 5 and Fig. 6.

Conclusions
The critical analysis of the DCB test configuration and direct applicability of the ASTM D5528 Standard in case of the mechanically coupled laminated composite beams was performed. The different initiation criteria was exploited to obtained the mode I critical strain energy release rate. Additionally to calculated the values of G_{IC} the CCM and the MBT methods were used. The results shows that the mode I c-SERR depends on specimen interface, mechanical coupling as well as the adopted initiation point.

Acknowledgments
The paper was financially supported by the Ministerial Research Project No. DEC-2016/21/B/ST8/03160 financed by the Polish National Science Centre.

References

[1] Syafigah Nur Azirie Safri et.al, Impact behaviour of hybrid composites for structural applications: A review, Compos. Part. B 133 (2018) 112-121.

[2] ASTM D 5528-01. Standard test method for mode I interlaminar fracture toughness of unidirectional fiber-reinforced polymer matrix composites (2001). https://doi.org/10.1520/d6671_d6671m-13

[3] A. B. Pereira, A. B. de Morais, Mode I interlaminar fracture of carbon/epoxy multidirectional laminates, Comp. Sci. Tech 64 (2004) 2261-70. https://doi.org/10.1016/j.compscitech.2004.03.001

[4] A. B. Pereira, A. B. de Morais,, Mode II interlaminar fracture of glass/epoxy multidirectional laminates, Compos. Part A 35(2) (2004) 265-72. https://doi.org/10.1016/j.compositesa.2003.09.028

[5] S. Samborski, Numerical analysis of the DCB test configuration applicability to mechanically coupled fiber reinforced laminated composite beams, Compos. Struct 152 (2016) 477-487. https://doi.org/10.1016/j.compstruct.2016.05.060

[6] S. Samborski, Analysis of the end-notched flexure test configuration applicability for mechanically coupled fiber reinforced composite laminaes, Compos. Struct 163 (2017) 342-349. https://doi.org/10.1016/j.compstruct.2016.12.051

[7] S. Samborski, Prediction of delamination front's advancement direction in the CFRP laminates with mechanical couplings subjected to different fracture toughness tests, Compos Struct 202 (2018) 643-650. https://doi.org/10.1016/j.compstruct.2018.03.045

[8] S. Samborski, J. Rzeczkowski, Numerical modeling and experimental testing of the DCB laminated composite beams with mechanical couplings, AIP. Conf. Proc 1922 (2018) 080010-1 – 6. https://doi.org/10.1063/1.5019081

[9] J. Rzeczkowski, S. Samborski, J. Paśnik, Experimental verification on the DCB test configuration applicability to mechanically coupled composite laminates, IOP. Conf. SER.: Mater. Sci. Eng, 416 (2018) 012055. https://doi.org/10.1088/1757-899x/416/1/012055

[10] J. Rzeczkowski, S. Samborski, J. Paśnik, Experimental investigation of mechanically coupled composite specimens in the ENF test configuration., IOP Conf. Ser.: Mater. Sci. Eng. 416, (2018) 012041. https://doi.org/10.1088/1757-899x/416/1/012041

[11] S. Samborski, J. Rzeczkowski, J. Paśnik, Issues of direct application of fracture toughness determination procedures to coupled composite laminates, IOP Conf. Ser.: Mater. Sci. Eng. 416 (2018) 012056. https://doi.org/10.1088/1757-899x/416/1/012056

[12] J. Paśnik, S. Samborski, J. Rzeczkowski, Application of the CZM technique to delamination analysis of coupled laminate beams, IOP Conf. Ser.: Mater. Sci. Eng. 416, (2018) 012075. https://doi.org/10.1088/1757-899x/416/1/012075

Experimental Mechanics of Solids
Materials Research Proceedings **12** (2019) 59-64

Materials Research Forum LLC
https://doi.org/10.21741/9781644900215-8

Experimental and Numerical Examination of the Fastening System's Rail Clip Type SB-4

Ludomir J. Jankowski[1,a *], Jaroslaw Filipiak[1,b], Malgorzata Zak[1,c]
Krzysztof Kruszyk[2,d]

[1] Wroclaw University of Science and Technology, Faculty of Mechanical Engineering,
50-370 Wroclaw, Poland

[2] Vossloh SKAMO Sp. z o.o., 63-460 Nowe Skalmierzyce, Poland

[a]ludomir.jankowski@pwr.edu.pl, [b]jaroslaw.filipiak@pwr.edu.pl,
[c]malgorzata.a.zak@pwr.edu.pl, [d]krzysztof.kruszyk@vossloh-skamo.pl

Keywords: Rail Fastening System, Spring Clip, Photoelastic Coating Technique, Strain Gauges Technique

Abstract. The paper presents the results of an experimental and numerical analysis of the effort of the elastic clip type SB-4 VK, during the process of its manual fastening on an anchor embedded in the pre-stressed concrete sleeper. The techniques of photoelastic coating and electrical resistance strain gauges were applied and a numerical simulation of this process was carried out to determine the level of the material effort of the attachment key component. The obtained results of the experiments and numerical calculations showed an acceptable level of the clip material effort.

Introduction

The use of elastic systems for fixing rails to railway track made of pre-stressed concrete sleepers is a solution that increases comfort and safety of driving. Systems such as, for example, the type W [1], e-Clip [2], FASTCLIP [3], as well as the SB system [4], are characterized by high vibration damping efficiency, while ensuring the necessary requirements resulting from applicable international standards [5], including the rail pressure to the ground. The basic element of such systems are elastic clips of different geometry. Due to the specific geometry, the embedding of clips in anchors fixed in sleepers generates complex stress states, hence the assessment of the state of effort during this, usually one-off, working phase is important for an assessment of the fatigue strength of the clamp.

The spring clip of the SB-4 VK fastening system is made of spring steel PN-EN 50S2 (1.5024) and is used to clamp the UIC60 (60E1) rail. The clip is attached to the pre-tensioned concrete sleepers using a suitably shaped anchor - Fig. 1a. The manual fastening procedure involves the use of the lever shown in Fig. 1b. It is also possible to remove the clip to change the force applied to the lever.

Experimental tests were carried out on the randomly selected clips provided by the manufacturer, after verifying (in accordance with company standards) the quality of the produced series of springs. The clips were embedded in the anchor by hand following the procedure provided by the manufacturer [6].

Experimental Mechanics of Solids
Materials Research Proceedings **12** (2019) 59-64

Materials Research Forum LLC
https://doi.org/10.21741/9781644900215-8

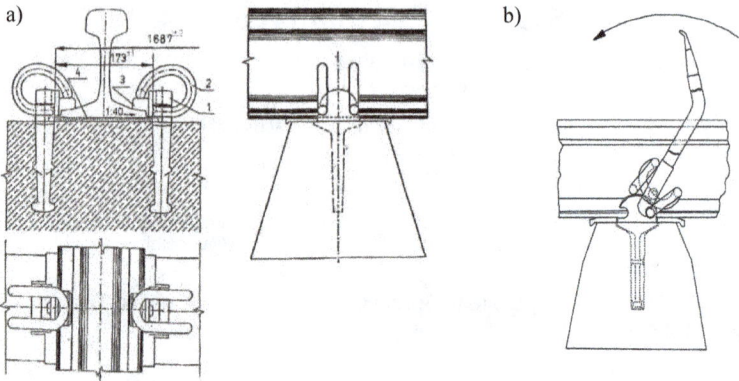

Fig. 1. SB fastening system of the UIC 60 rail [6]: a) 1 – anchor, 2 – spring clip, 3 – electro-insulating pressure pad, 4 – elastic rail pad, b) scheme of the SB-4 VK clip manual deposition

Experimental investigations

Basing on the preliminary analysis of the clip's work during its fastening, a decision was made to use the photoelastic coating technique [7,8] and the resistance strain gauge method [7,8] to measure the strains during the process characterized by the largest deformation of the geometry of this attachment element. The tested clips were fastened, according to the assembly instructions, on a system provided by the manufacturer simulating fixing the rail to the underlay. Practically, with one exception of the concrete railway sleeper, all elements of the SB4 VK fastening system and their geometry meet the conditions specified in EN 13481-2 and TSI (Technical Specification for Interoperability) [5].

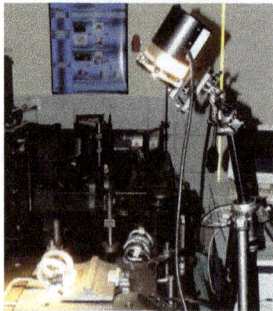

Fig 2. View of the measurement system

Fig. 3. SB4 clips on a stand

The view of the measurement system during the measurement with the photoelastic coating technique is shown in Fig. 2, and the view of the rail fastening system with the clip covered with photoelastic coating - in Fig. 3. Due to the application of the optical strain measurement method, only the UIC 60 rail bottom part was affixed, which facilitated the observation of optical effects in the layer glued to the previously cleaned surface of the clip. In the case of the discussed measurements, a coating of $t_c = 2.0$ *mm* thickness was used. Due to the shape of the examined surfaces of the clip, the coating was made in the form of so-called contour plates.

A composition based on an epoxy resin was used as the coating material. The plates formed on the surface of the clip were glued to the surfaces to be tested, using a reflective glue prepared on the basis of the above-mentioned composition with an addition of aluminum dust. For the purpose of quantitative analysis of the state of deformation, the value of isochromatic fringe pattern was determined, as $f_\varepsilon = 1.58 \times 10^{-3} [-]$.

The measurements were carried out using a V-type reflection polariscope - Model 031, by Photoelastic Vishay Inc., recording photoelastic images with a digital camera. Due to the dynamics of the clip fastening process, the video mode was the basic recording mode. Sample frames of the registered changes of the image of full-order isochromatic fringes during the process of fastening as a function of standardized time T from the interval [0,1] are presented below in Fig. 4.

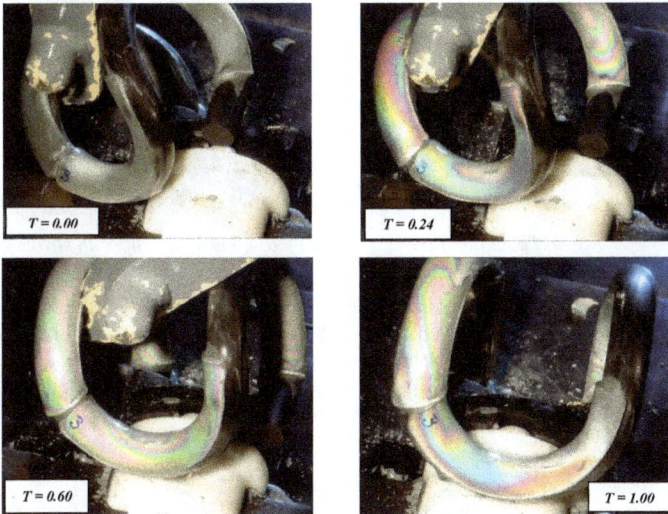

Fig. 4. Views of the full-order isochromatic fringes during fastening SB4 clip

In this case of photoelastic coating technique research, the isochromatic fringe corresponds to the difference of the principal strains:

$$\varepsilon_1 - \varepsilon_2 = N f_\varepsilon \tag{1}$$

where: N – isochromatic fringe order, f_ε – strain-value of the isochromatic fringe.

Due to the dominance of the deformations caused by the bending of individual fragments of the clip, a correction factor [7,8] was introduced, which captures the influence of the strain gradient on the coating thickness:

$$(\varepsilon_1 - \varepsilon_2)^o = (\varepsilon_1 - \varepsilon_2)^c \cdot K_b = N f_\varepsilon K_b \tag{2}$$

Experimental Mechanics of Solids   Materials Research Forum LLC
Materials Research Proceedings **12** (2019) 59-64   https://doi.org/10.21741/9781644900215-8

In the case under consideration, the value $of K_b$ is: $K_b = 0.897$.

The difference of the principal stresses, for $E^o = 2.05 \times 10^5 MPa$ and $v^o = 0.3$, is determined by the equation:

$$(\sigma_1 - \sigma_2)^o = \frac{E^o}{(1+v^o)} N f_\varepsilon K_b = 141.45 \cdot N \ [MPa]$$

(3)

While fastening the clip, the maximum values of the isochromatic fringe order range of $N \cong 6.5$ were observed, which corresponds to the difference of the principal stresses of 920 MPa.

The preliminary, qualitative analysis of the registered images of isochromatic fringes enabled the identification of the locations of the resistance strain-gauges, which is shown in Fig. 5.

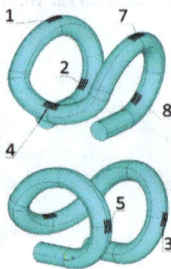

Fig. 5. Location of the

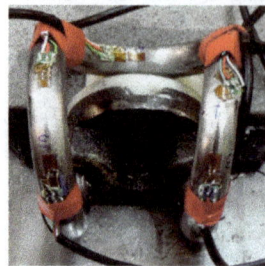

Fig. 6. View of the clip with strain-gauges resistance strain-gauges

Foil strain gauges with a measuring base a = 3 mm were used. They were glued in selected places in the Wheatstone half-bridge system [8]. The measurements were carried out with the Stretton 0316E measurement system allowing simultaneous sampling of 16 measurement channels. The view of the clip with glued gauges is shown in Fig. 6. An example of the changes registered in the stress value during the clip fastening is presented in Fig. 7.

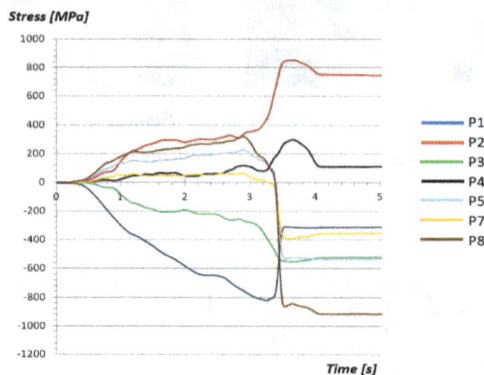

Fig. 7. Changes of the normal stress in the selected (P1-P8) points

Experimental Mechanics of Solids   Materials Research Forum LLC
Materials Research Proceedings **12** (2019) 59-64   https://doi.org/10.21741/9781644900215-8

Numerical simulations

Numerical simulations of the stress state in the spring clip SB4 were carried out using the finite element method (FEM) in the ANSYS Workbench 16.2 environment. The geometric model of the SB4 clip and anchor (excluding its stem) is shown in Fig. 8.

The tests included both static and dynamic simulations, allowing to determine the distribution of stresses for the following cases:

a) bending (stretching) the free arms of the SB4 clip with a force of 6.8 kN,

b) operation with a bending force of 14 kN on the central arc of the SB4 clip,

c) fastening the SB4 clip on the anchor.

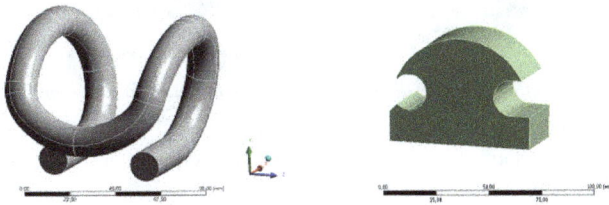

Fig. 8. Geometric models of the clip and anchor head

The following material parameters of spring steel 50S2 (1.5024) were adopted for numerical calculations: density $\rho = 7850\ kgm^{-3}$, Young's modulus $E = 205\ GPa$, Poisson's ratio $v = 0.3$, elastic limit $R_e = 1250\ N/mm^2$.

The static simulations carried out for standard clip loads (cases a) and b)) showed permissible values of maximum stresses, i.e. $\sigma_{max} < R_e$ [9].

In case of c) a simulation of clip fastening on the anchor was carried out, using the software for the analysis of fast-changing phenomena employing the LS-DYNA finite element method. The geometric model of the initial position of the clip SB4 with respect to the anchor is shown in Fig. 9.

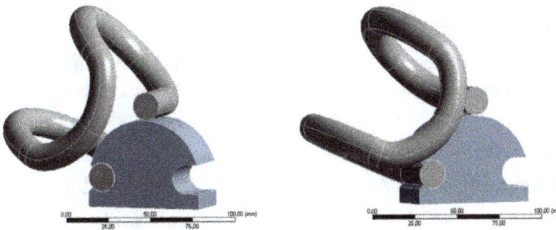

Fig. 9. Geometric model of the initial position of the clip (for dynamic simulation of the clip fastening)

An example of the distribution of the stress difference contours $(\sigma_y - \sigma_z)$, obtained in the final stage of fastening the clip, is shown in Fig. 10a, while Figure 10b shows the contours of the reduced stress (von Mises) for the same clip position.

Materials Research Forum LLC

https://doi.org/10.21741/9781644900215-8

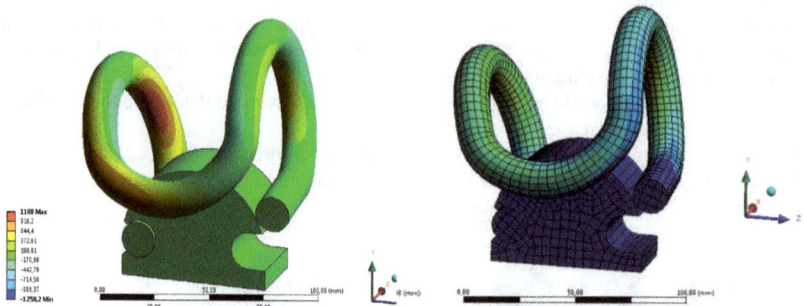

Fig. 10. Contours of the: a) stress difference $(\sigma_y - \sigma_z)$ *in final clip fastening stage, b) reduced stress* σ_{eq} *(von Misesa)*

The simulations performed confirmed the acceptable level of effort of the SB4 clip material during its fastening: $\sigma_{eq} < R_e$.

Summary

The above-mentioned experimental tests and numerical calculations were performed for the critical load case of the SB4 clip. The obtained results confirmed the acceptable level of effort of the clip material during its fastening, with the occurrence of complex states of temporarily high stress states being the characteristic feature of this process.

References

[1] Information on http://www.pandrol.com/application/high-speed/

[2] Information on http:// www.railway-fasteners.com

[3] Information on http:// www.pandrol.com/product/pandrol-fastclip-fd/

[4] Information on http:// www.vossloh.com/en/

[5] Information on https://publications.europa.eu, Commission Regulation (EU) No 1299/2014 of 18 November 2014 on the technical specifications for interoperability relating to the "infrastructure" subsystem of the rail system in the European Union

[6] Instructional material provided by Vossloh (in polish). (2016)

[7] J.W. Dally, Experimental Stress Analysis, third ed., McGraw-Hill, Inc., New York, 1991.

[8] A. S. Khan, X.Wang, Strain Measurements and Stress Analysis, first ed., Prentice Hall, 2001.

Experimental Mechanics of Solids
Materials Research Proceedings **12** (2019) 65-70

Materials Research Forum LLC
https://doi.org/10.21741/9781644900215-9

Application of the Photoelastic Coating Technique in Car Suspension Failure Analysis

Ludomir J. Jankowski[1,a *], Marek Reksa[1,b],

[1] Wroclaw University of Science and Technology, Faculty of Mechanical Engineering,
50-370 Wroclaw, Poland

[a]ludomir.jankowski@pwr.edu.pl, [b]marek.reksa@pwr.edu.pl

Keywords: Trailing Arm Failure, Photoelastic Coating Technique, Fatique Fracture

Abstract. The article presents the results of an experiment carried out to determine the cause of damage to the right rear trailing arm of a passenger car. The routine inspection of the vehicle and the test of the arm damaged during the running have indicated production defects as the cause of a dangerous failure. The results of the expertise were questioned and the thesis was formulated that the process of the rocker failure was initiated during the theft of wheels by raising the car in a manner inconsistent with the manufacturer's instructions. A simulation of this event was carried out by lifting the vehicle on the trailing arm. To assess the effort of the rocker, the photoelastic coating technique was used. The analysis of the measurement results showed that the theft incident could not initiate the fatigue process that led to the failure of the trailing arm.

Introduction

When driving a van car with normal operational load, on the motorway at the permitted speed, the driver of the vehicle began to feel the rear axle swimming, which indicated losing the stability of moving. Despite the driver's attempt to control this phenomenon, the car became unsteady after a while, which led to hitting the protective barrier, performing several revolutions around the center of gravity of the vehicle and finally stopping on the road.

In the insurance indemnification procedure, a visual inspection of the damaged elements was carried out routinely, which confirmed the course of the accident described by the driver, and also showed a cause and effect relationship between the damage to the right longitudinal rear trailing arm of the car and the incident. In particular, the fact that the trailing arm failed while driving the car with a relatively small mileage (about 37000 km), led to metallographic expertise of the damaged part of the suspension [1]. It pointed to production defects as the source of fatigue processes initiation, which resulted in the abovementioned trailing arm failure.

As an incident of a theft of wheels was registered in the vehicle history, the manufacturer's representative questioned the results of the expert opinion and indicated the plastic strain occurring while lifting the car *on the rocker* as the primary source of its damage. Thus, in order to experimentally verify this thesis, it was decided to conduct an experimental simulation of the theft of the wheel and determine the values of the stress and the places of its concentration.

The object and measurement method

The tests were carried out on the trailing arm of a brand new rear suspension beam of the same car model. It was decided to use a surface method for measuring strains, i.e. a photoelastic coating technique [2,3]. It allows to evaluate the state of strain and stress on the surface of an object covered with a thin layer of material exhibiting the effect of birefringence. Optical effects observed in polarized light are related to the parameters of the strain field (and in the linear-elastic range - with stress), which allows for qualitative and quantitative analysis.

Experimental Mechanics of Solids
Materials Research Proceedings **12** (2019) 65-70

Materials Research Forum LLC
https://doi.org/10.21741/9781644900215-9

The basic relation defining the relationship between the optical effect (so-called isochromatic fringe order) and the principal strain difference is:

$$\varepsilon_1 - \varepsilon_2 = N \cdot f_\varepsilon$$
(1)

where: N – isochromatic fringe order, $(\varepsilon_1 - \varepsilon_2)$ - difference of the principal strains, f_ε – strain-value of the coating.

Due to the shape of the swingarm surface, the photoelastic coating in the form of so-called contour plate with a thickness of $t_c = 2.5\ mm$, was made of epoxy resin and then bonded with a reflective glue. The rear axle view, with the visible coating glued on the arm, is shown in Figure 1.

Fig. 1. View of the photoelastic coating on the trailing arm's side surface
The rear beam prepared in this way was mounted in the same, technically efficient car model.

The measurements procedure

In the discussed experiment, the simulation of lifting the car in a manner inconsistent with the operating instructions, was carried out in the polygon conditions. By using a telescopic screw jack acting directly on the trailing arm (in a place defined by the wheel outline), the right rear wheel was lifted and the photoelastic effects were recorded. At the same time, the height of the selected body point was measured. The measurements started from state A, which corresponded to the minimum lift height of the rocker enabling the wheel removal, and ended by resting the rocker on bricks (state E).

The measurements were carried out using a Vishay reflective polariscope (model 031) and a Sony digital camera. The images of full- and half-order isochromatic fringes were recorded, and also the isochromatic fringe order was compensated at selected points of the photoelastic coating. The photoelastic images for individual lifting phases are shown below.

Fig. 2. Full-order isochromatic fringes – lifting phase A($N_{max} \cong 1.1$)

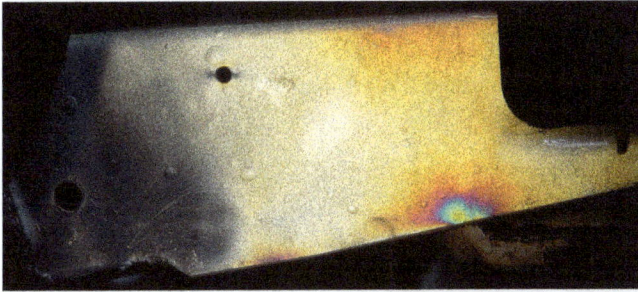

Fig. 3. Full-order isochromatic fringes – lifting phase B ($N_{max} \cong 1.5$)

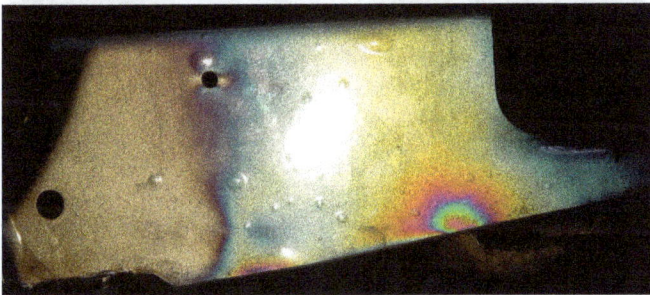

Fig. 4. Half-order isochromatic fringes – lifting phase C ($N_{max} = 2.35$)

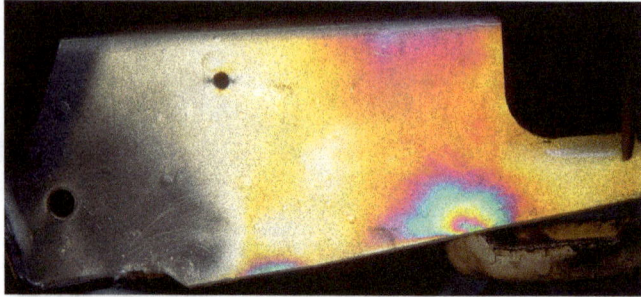

Fig. 5. Full-order isochromatic fringes – lifting phase D ($N_{max} \cong 2.0$)

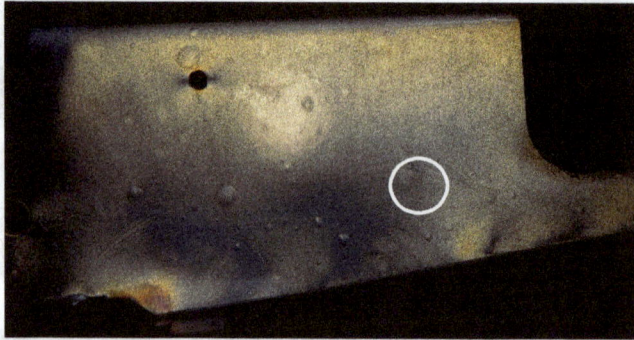

Fig. 6. Full-order isochromatic fringes – phase E (supporting on the bricks) ($N_{circle} \cong 0.3$)

Analysis and results

When analyzing registered isochromatic images, the location of maximum isochromatic fringe order N was compared with the fracture line in the rocker damaged during the accident - Fig. 7a. This fracture runs through the hole for fixing the handbrake cable, approximately perpendicularly to the upper edge of the rocker, and it is located in the area covered by the wheel(!). Fig. 7b shows its course from the inside in a sample cut for the metallographic research. As it can be seen, the fracture also runs through the end of the weld fixing the reinforcement (rib) to the rocker.

In contrast, during the wheel theft simulation, the maximum values of the isochromatic fringes occurred in the place of impact of the jack head on the lower edge of the trailing arm, away from the fracture.

Experimental Mechanics of Solids
Materials Research Proceedings 12 (2019) 65-70

Materials Research Forum LLC
https://doi.org/10.21741/9781644900215-9

Fig. 7. View of the fracture line: a) from the outer side of the arm, b) from the inside (view rotated 90 ° to the left, Sz - main weld, Sw - preliminary weld, H1, H2 - designation of sample fragments)

Because the point impact of the jack head is practically the most unfavorable case of the suspension arm load (in order to enable a wheel removal), it was assumed that the quantitative analysis of photoelastic images would be limited to estimating the stress value acting tangentially to its lower edge, in the immediate vicinity of the jack head.

For this purpose, the so-called the stress value of the isochromatic fringe f_σ was calculated:

$$f_\sigma = \frac{E_o}{1+v_o} f_\varepsilon \tag{2}$$

where: E_o – object's modulus of elasticity ($E_o = 2.05 \times 10^5$ MN/m^2), v_o - object's Poisson's ratio ($v_o = 0.3$), f_ε - strain-fringe value (in the case under consideration $f_\varepsilon = 1.105 \times 10^{-3}$). The value of normal stress acting tangentially to the edge of the rocker is determined by the formula:

$$\sigma_1 = N f_\sigma \tag{3}$$

The maximum stress values for lifting phases are given in Table 1.

Table 1. Maximum stress at the lower edge of the trailing arm

Lifting phase	Height of the car body [mm]	σ_{max} [N/mm^2]
A	258	191.7
B	265	261.4
C	270	409.5
D	275	348.5
E	support on the bricks	174.3

During the lifting, an increase in the stress value was initially observed (max approx. 410 N /mm^2 in phase C), followed by its decrease due to the inclination of the body and the shift of the vehicle center of gravity to the left (phase D). After removal of the jack and support *on bricks*, a low isochromatic fringe order ($N \approx 0.3$) was observed at the location of the jack head impact (area marked with a circle in Fig. 6), so the yield point was exceeded in this region; however, the stress corresponding to the plastic strain reached only approx. 50 N/mm^2.

Summary

The simulation of lifting the vehicle on the rear arm excludes the possibility of fatigue processes being initiated by this incident. A large distance of the fracture line from the region in which the

Experimental Mechanics of Solids   Materials Research Forum LLC
Materials Research Proceedings **12** (2019) 65-70   https://doi.org/10.21741/9781644900215-9

maximum stress values were observed (and after the removal of the jack – low value of the plastic deformation), confirms this conclusion.

It should be emphasized that the metallographic expert opinion [1] clearly indicated the focus of the crack in the joint connecting the reinforcing rib to the arm. Other findings included:

- the ending of the connecting joint (8 mm) and the main joint (20 mm) before the end of the reinforcing rib, visible in Fig. 6b,
- no weld penetration (Fig. 7),
- gas bubble in the weld and non-metallic inclusions originating from the paint layer (not removed before welding - Fig.8).

Fig. 7. No penetration in the joint connecting the rib to the trailing arm

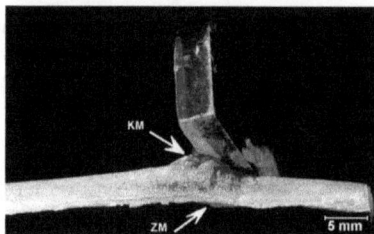

Fig. 8. The view of the fracture in the rib's area - the lack of penetration and not removed varnish layer (KM - fragile fracture zone, ZM - fatigue zone)

Each of these defects causes stress concentration, which in combination with a variable load resulted in the formation of fatigue-type fracture with a ductile fracture zone.

Therefore, the failure is not a result of an improper operation and it was not caused by the user of the vehicle.

References

[1] W. Dudziński, M. Lachowicz, Metallographic examination of car axle components after an accident (in polish), unpublished expert opinion, Wroclaw University of Science and Technology (2004)

[2] J.W. Dally, Experimental Stress Analysis, third ed., McGraw-Hill, Inc., New York, 1991.

[3] A. S. Khan, X. Wang, Strain Measurements and Stress Analysis, first ed., Prentice Hall, 2001.

Experimental Mechanics of Solids
Materials Research Proceedings **12** (2019) 71-76

Materials Research Forum LLC
https://doi.org/10.21741/9781644900215-10

Local Plastic Instabilities of Perforated Thin-Walled Bars – FEM Modelling and DIC Verification

Andrzej Piotrowski[1,a*], Marcin Gajewski[1,b] and Cezary Ajdukiewicz[1,c]

[1]Warsaw University of Technology, Civil Engineering Department, Armii Ludowej 16 Street, 00-637 Warsaw, Poland

[a]a.piotrowski@il.pw.edu.pl, [b]m.gajewski@il.pw.edu.pl, [c]c.ajdukiewicz@il.pw.edu.pl

Keywords: Finite Element Method Modelling, Digital Image Correlation, Buckling, Critical Forces, Post-Critical Modes, Thin-Walled Bars, Perforated Bars

Abstract. In this paper results of testing and modelling of perforated thin-walled bars of low slenderness are shown. Tested and modelled elements in general are used as structural elements for storage systems. In such case the compression mode is dominant, so the proper understanding of the element behaviour in post-critical stage is essential for system safety estimation. To predict post critical behaviour the suitable constitutive models of elasto-plasticity are needed. The most essential to such simulations is the fact that local deformations and rotation angles are significant, so the large deformation modelling regarding to geometry and constitutive models have to be used. In the paper for FEM (finite element method) modelling the ABAQUS system is used, and obtained solutions are verified experimentally using Instron 8802 universal testing machine. Aside from measuring critical forces and final deformations for several samples and eight different bar's lengths, also strain and displacement fields were verified with application of DIC (digital image correlation) system ARAMIS. Because of the testing machine and DIC system limitations only very low slenderness (1÷11) samples were taken into account. Bars of such a low slenderness should be treated as shells or three dimensional objects in numerical modelling.

Introduction

Storage systems are one of the most popular applications of perforated thin-walled bars, but the design rules for such systems can give results on the unsafe side [1], leading to many dangerous situations (structure failures) and financial losses. The paper presents results of experiments and numerical modelling of compression tests of elements presented in Fig.1. These elements are taken from a real storage system partially destroyed because of the accumulated damage created by the light impacts with the forklifts operating in the magazine area. It is the continuation of work presented on XXII, XXIII and XXIV LSCE conferences: thin-walled bars theory [2-6] calculations as well as finite element method modelling [4, 7-10] were presented in paper [11], elastic buckling modes were analysed in paper [12] and experimental results can be found in paper [13]. Some very early experiments and calculations were also presented in paper [14]. Paper [15] presents some experimental results, especially displacement fields registered with ARAMIS system. In the previous papers only some small, then studied parts of research were presented. This paper gives the general, less detailed, but more complete view on the project. It shortly presents and compares most recent and complete results obtained by the various methods.

Description of elements used for numerical and experimental tests

Elements chosen for testing are presented on Fig. 1 and are made from a typical low carbon steel [13]. They were fixed on both ends and compressed statically without any eccentricity. Lengths

Experimental Mechanics of Solids
Materials Research Proceedings **12** (2019) 71-76

Materials Research Forum LLC
https://doi.org/10.21741/9781644900215-10

of the tested samples were 50, 100, 150, 200, 250, 300, 400 and 500 mm, which corresponds to slenderness varying from 1 to 11.

For the analysis presented in this paper eight finite element method models were created in ABAQUS software. Each model consists of shell elements – mainly S4R but also a small number of S3 elements. Both kinds are elements with linear shape functions and reduced integration, S4R – quadrangular and S3 – triangular. Mesh was created using the automatic ABAQUS' module and the triangular elements are used near the perforations, as it would be much more difficult, if even possible, to use only quadrangular elements in these places.

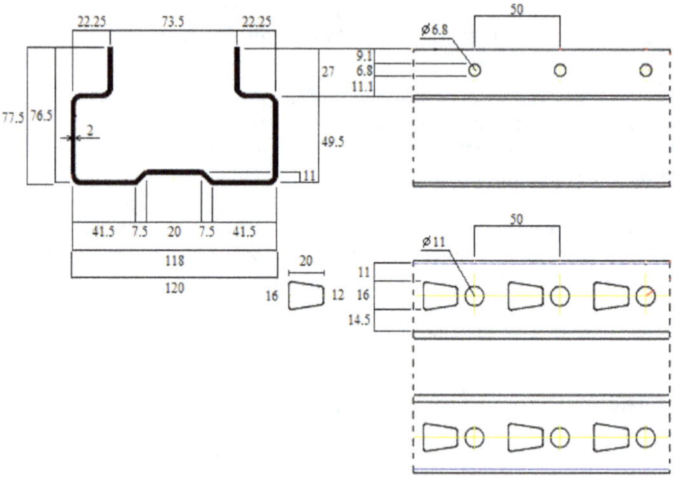

Fig. 1. Characteristic dimensions of tested elements [14].

Approximate size for the global seed was 2 mm. Table 1 presents how many elements of each type were used in the models. The displacement type boundary conditions were applied to the nodes lying at the ends of the modelled bars. For the fixed boundary conditions all of the DOFs were blocked. One end of the bar was fully fixed and displacement causing the compression was applied to the nodes lying at the other end of the bar.

Table 1. Numbers of elements used in the models:

Element type	Number of elements for models of length equal to:							
	50 mm	100 mm	150 mm	200 mm	250 mm	300 mm	400 mm	500 mm
S4R	3946	7794	11681	15569	19448	23337	31125	38938
S3	42	82	140	170	218	244	340	440

The quasi-static displacement-controlled analysis by ABAQUS/Standard with large deformation theory (NLGEOM option) was used. Solution technique was full Newton and time increment was automatically adjusted (maximum number of increments was 1000, initial increment size was 0.01, and minimum increment size was 1E-007).

Experimental Mechanics of Solids   Materials Research Forum LLC
Materials Research Proceedings **12** (2019) 71-76   https://doi.org/10.21741/9781644900215-10

The analysed bars were made from low carbon steel [15]. It is treated as an isotropic material and described through elasto-plastic constitutive model. In elasticity range material is characterized as linear with Young modulus equal to 230 GPa and Poisson ratio equal to 0.3. Material becomes plastic after crossing Huber-Mises yield criterion. The permanent deformation is calculated on the basis of associated with yield condition flow rule leading to plastic part of the logarithmic strain tensor [7]. Strain hardening/softening is assumed as isotropic one. The parameters for both elasticity and plasticity were determined based on laboratory experiments. For application of the large deformation theory implemented in ABAQUS proper material data preparation is needed [16]. Therefore, Piola stress axial component was recalculated onto Cauchy stress and nominal strain was recalculated onto logarithmic strain tensor axial component [7].

Description of experimental tests and obtained results
Tested samples were mounted in Instron 8802 universal testing machine using compressing plates with 5 mm fixtures (see Fig. 2). They were compressed with the displacement speed calculated as 4% of initial length per minute and the test was ended, when the distance between compression plates was equal to the 90% of its initial value. Force and displacement were recorded during the test with the testing machine, and the chosen permanent displacements were measured manually after the test ended [13]. Whole process was also analyzed [15] using ARAMIS digital image correlation system [17].

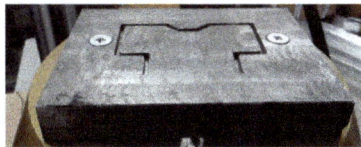

Fig. 2. Compressing plates with fixtures [13].

Three buckling modes were observed (see Fig. 3.). In the experiments the first (fully local) mode was characteristic for the shortest elements (50 mm and 100 mm length), the second mode for the elements longer than 150 mm and the third mode was possible to obtain in the longest (500 mm long) samples through grinding their ends and in the result – adding some eccentricity. 150 mm long samples are the border between two main buckling modes and because of that can be destroyed in the many ways combining both of them. In the numerical modelling the first mode was also characteristic for the shortest elements (50 mm and 100 mm length) and from 150 mm onwards both second and third mode were possible in every case.

First mode: Second mode: Third mode:

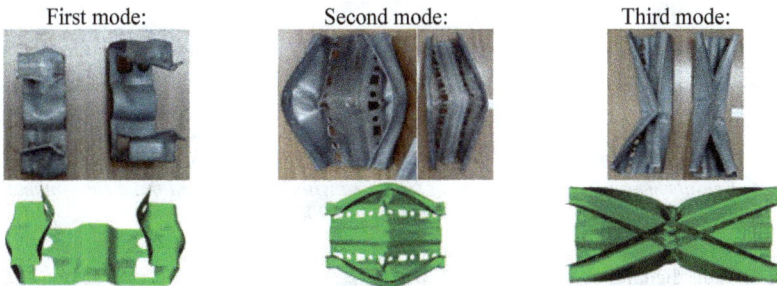

Fig. 3. Buckling modes.

Comparison of Huber-Mises strains acquired from experiment and Huber-Mises stresses acquired from FEM modelling shows good quality compliance (see Fig. 4.).

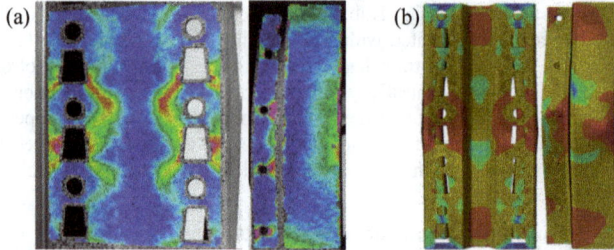

Fig. 4. Huber-Mises strain field from Aramis DIC system (a) and Huber-Mises stress field from Abaqus FEM software (b) for the 250 mm long sample.

Equilibrium paths acquired from experiment and calculations for two chosen lengths are shown in Fig. 5. Experimental data are presented without any processing, especially initial fitting stages are not cut from the graphs.

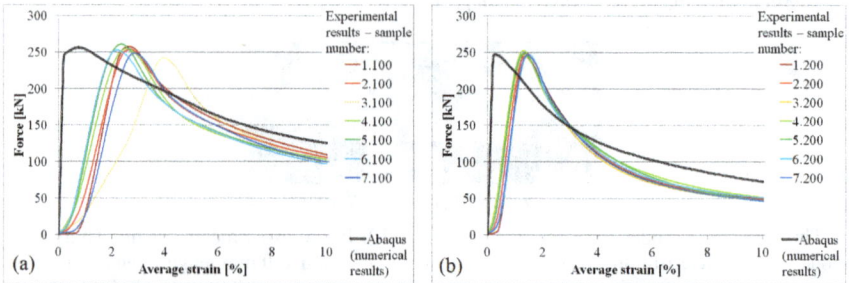

Fig. 5. Equilibrium paths from experiment (without DICS or extensometers) and finite element method modelling for 100 mm (a) and 200 mm (b) bars.

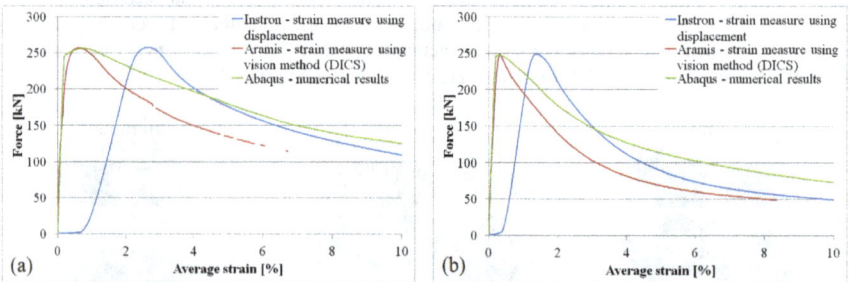

Fig. 6. Equilibrium paths from experiment (with and without DICS) and finite element method modelling for 100 mm (a) and 200 mm (b) bars.

Using Aramis digital image correlation system for acquiring strains instead of evaluating it from machine data gives better results and better compliance with theoretical results (see Fig. 6.).

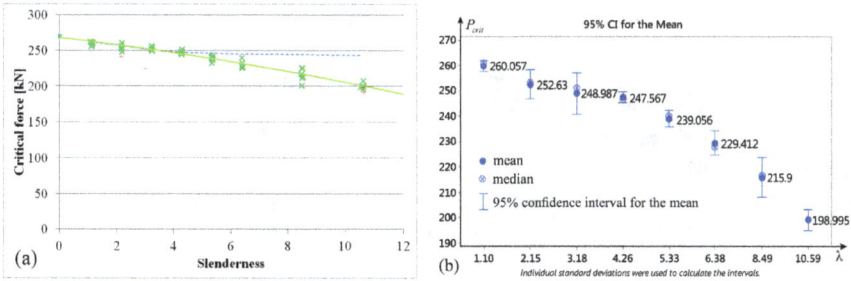

Fig. 7. Critical force in the function of slenderness: green – experiment, blue – FEM calculations (a), and statistical analysis of experimental results (b).

Critical forces, understood as maximum forces acquired during the tests, are correctly predicted for elements shorter or equal 20 cm. For longer elements calculations give higher values than observed in reality (see Fig. 7.).

Summary
Comparison of numerical and experimental results shows good quality compliance. Buckling modes are correctly predicted (Fig. 3.) and areas of the biggest strains are also identified properly (Fig. 4.). Formation of the curved fields of high strains connecting perforations can be observed as the most important part of the buckling process, especially in the shortest samples, in which the buckling is plastic and local. Critical forces are correctly predicted only for the shortest elements (up to 20 cm of length – see Fig. 7.). Possible reasons for overestimation of this value in the longer ones are boundary conditions (fixed in theory and in reality realised as shown on Fig. 2.) and imperfections (thin-walled bars can be very sensitive to even the smallest differences between theoretical and real geometry [18]). Elastic phase of compression is properly modelled, in the plastic phase some differences between model and experiment can be observed (Fig. 6.). When comparing theoretical and experimental results acquiring correct strains from experiment is very important. It can be done by using extensometers or digital image correlation system. Ignoring this can result in big errors because of the problem of contact between tested samples and compression plates (see Fig. 5.). Further work is necessary: better modelling of fixed boundary conditions, testing elements with hinged boundary conditions, investigating imperfections, trying different material models and better determination of parameters for them (for example reverse analysis).

References

[1] C. Bernuzzi, F. Maxenti, European alternatives to design perforated thin-walled cold-formed beam–columns for steel storage systems, J CONSTR STEEL RES, 110 (2015) 121–136. https://doi.org/10.1016/j.jcsr.2015.02.021

[2] M. Kotelko, Load capacity and mechanisms of destruction of thin-walled structures, Wydawnictwo Naukowo-Techniczne, Warsaw, 2011 (in Polish).

[3] J. Mutermilch, A. Kociolek, Strength and stability of thin-walled bars with open cross-section. Warsaw University of Technology Publishing House, Warsaw, 1964 (in Polish).

[4] M. Nedelcu, Buckling mode identification of perforated thin-walled members by using GBT and shell FEA, Thin-Walled Structures 82 (2014) 67–81. https://doi.org/10.1016/j.tws.2014.04.005

[5] J.B. Obrebski, Thin-walled elastic bars, Warsaw University of Technology Publishing House, Warsaw, 1991 (in Polish).

[6] S. Piechnik, Thin-walled bars with open cross-section, Cracow University of Technology Publishing House, Cracow, 2000 (in Polish).

[7] ABAQUS User's manual, ver. 6.11. Dassault Systèmes, SIMULIA (2011) 130.149.89.49:2080/v6.11/ (access on 08-03-2019).

[8] G. Rakowski, Z. Kacprzyk, Finite element method in structural mechanics, Warsaw University of Technology Publishing House, Warsaw, 2016 (in Polish).

[9] O.C. Zienkiewicz, R.L. Taylor, The finite element method for solid and structural mechanics, Sixth edition, Elsevier Butterworth Heinemann, 2005.

[10] O.C. Zienkiewicz, R.L. Taylor, J.Z. Zhu, The finite element method – Its basis & fundamentals, Sixth edition, Elsevier Butterworth Heinemann, 2005.

[11] L. Kowalewski, A. Piotrowski, M. Gajewski, S. Jemiolo, FEM application for determination of post-critical deformation modes of perforated thin-walled bars, in: Monograph from Scientific Seminar Organized by Polish Chapters of International Association for Shell and Spatial Structures, University of Warmia and Mazury, Faculty of Geodesy, Geospatial and Civil Engineering, XXII LSCE – 2016, Olsztyn, 2016, pp. 27-30.

[12] L. Kowalewski, A. Piotrowski, M. Gajewski, S. Jemiolo, Determination of critical forces with corresponding deformation modes for perforated thin-walled bars, in: Monograph from Scientific Seminar Organized by Polish Chapters of International Association for Shell and Spatial Structures, University Science and Technology, Faculty of Civil Engineering, Architecture and Environmental Engineering, XXIII LSCE – 2017, Bydgoszcz, 2017, pp. 14-17.

[13] A. Piotrowski, M. Gajewski, C. Ajdukiewicz, L. Kowalewski, S. Jemiolo, Experimental and numerical determination of critical forces for perforated thin walled bars, in: Monograph from Scientific Conference of IASS Polish Chapters, Lodz University of Technology, XXIV LSCE – 2018, Lodz, 2018, pp. 109-114.

[14] A. Piotrowski, L. Kowalewski, R. Szczerba, M. Gajewski, S. Jemiolo, Buckling resistance assessment of thin-walled open section element under pure compression, MATEC Web of Conferences, Vol. 86, Article Number: 01021, 2016. https://doi.org/10.1051/matecconf/20168601021

[15] A. Piotrowski, M. Gajewski, C. Ajdukiewicz, Application of digital image correlation system for analysis of local plastic instabilities of perforated thin-walled bars, MATEC Web of Conferences, Vol. 196, Article Number: 01032, 2018. https://doi.org/10.1051/matecconf/201819601032

[16] S. Jemiolo, M. Gajewski: Hyperelasto-plasticity. Warsaw University of Technology Publishing House, Warsaw 2017 (in Polish).

[17] Aramis v6.1 User Manual, GOM Gmbh (2008) materials-science.phys.rug.nl/index.php/home/downloads/category/1-manuals?download=27%3Aaramis-v61.J (access on 08-03-2019).

[18] P. Paczos, J. Kasprzak, Influence of actual imperfections on the strength and stability of cold-formed thin-walled C-beam, 28th Symposium of Experimental Mechanics of Solids, Jachranka, 2018.

Experimental Mechanics of Solids
Materials Research Proceedings **12** (2019) 77-83

Materials Research Forum LLC
https://doi.org/10.21741/9781644900215-11

Fatigue Crack Growth Rate in Long Term Operated 19th Century Puddle Iron

Grzegorz Lesiuk[1, a *], Jose A.F.O. Correia [2,b], Michał Smolnicki[1,c],
Abilio M.P. De Jesus[2,d], Monika Duda[1,e], Pedro Montenegro[2,f],
Rui A.B. Calcada[2,g]

[1] Wroclaw University of Science and Technology, Faculty of Mechanical Engineering, Department of Mechanics, Materials Science and Engineering, Smoluchowskiego 25, 50-370 Wrocław, +48 71 3203919, POLAND

[2] Faculty of Engineering, University of Porto, Rua Dr. Roberto Frias, 4200-465 Porto, (+351) 966559442, Portugal

[a] grzegorz.lesiuk@pwr.edu.pl, [b] jacorreia@fe.up.pt, [c] michal.smolnicki@pwr.edu.pl, [d] ajesus@fe.up.pt, [e] monika.duda@pwr.edu.pl, [f] paires@fe.up.pt, [g] ruiabc@fe.up.pt

Keywords: Puddle Iron, Fatigue Crack Growth, Crack Closure, Strain Energy Parameter

Abstract. The paper presents the results of an experimental investigation of the fatigue crack growth in plane specimens made from puddle iron. Eiffel Bridge from Portugal and 19th-century viaduct steel member from Poland. The tests were performed under the load ratios $R = 0.05, 0.1$ and 0.5. There were also considered the different description of fatigue crack growth rate using strain energy density parameter based on cyclic J-integral. The fatigue crack growth rate in the tested material is significantly higher than its "ancient" equivalent i.e. old 19th-century mild steel. There is also a noticeable strong contribution of the crack closure effect.

Introduction

It is essential to study the structural integrity of old metallic bridges in order to understand their limitations regarding both material and structure [1]. One aspect to consider is the fatigue behaviour of riveted structural details and materials. The metal bridges of the late 19th and early 20th centuries were built with metallic materials, nowadays called puddled steel or iron [1,2]. These properties are essential for residual life analysis of metallic bridges in order to extend the lifetime of these structures. The metallic material under study was obtained from beam elements extracted from the Eiffel riveted metallic bridge, designed by Gustavo Eiffel and inaugurated on June 30, 1878. This bridge is located in Viana do Castelo, Portugal, crossing the Lima River, having a total length of 645 m. Another part of an ancient iron member was extracted from the restored viaduct (1863) located in Brochocin (Low Silesia, Poland),. Due to limited knowledge of the fatigue crack growth behaviour [3], this paper fills the gap in the literature on experimental results of fatigue crack growth (FCG) behaviour based on the effective stress intensity factor range and of a new parameter using the strain energy density concept for the metallic materials under consideration [4]. These FCG results are of great importance, since it allows the residual lifetime assessment of old metallic bridges of the same historical period, reducing the uncertainties associated with this type of materials.

Material Investigation

This work is focused on the chemical analysis (C, Mn, Si, P, S) of the tested ancient material using spectra method (measurement station presented in Table 1 results represent the mean value of three measurements) and the microstructural analysis in the post-operated state. The

Experimental Mechanics of Solids   Materials Research Forum LLC
Materials Research Proceedings **12** (2019) 77-83   https://doi.org/10.21741/9781644900215-11

metallographic inspection was performed at magnification range from 100× to 1000× on non-etched and etched state of the sample (5%HNO₃) using the metallographic microscope following the ASTM E407 - 07(2015)e1 [5]. Images were registered by coupled with microscope digital camera CMOS 15MPx with the software for image analysis.

Table 1. The chemical composition of tested materials (in % by weight)

Investigated materials	C	Mn	Si	P	S
Eiffel Bridge	<0.01	0.01	0.07	0.354	0.045
Puddle iron 1863 [2]	0.08	0.025	0.15	0.245	0.015

Based only on the chemical analysis there is noticeable high phosphorus content in both materials. This fact indicates for possible classification of the tested materials as a puddle iron. Based on the theory of degradation of old puddle iron [6,7], the low carbon content (<0.1%C) and lack of the de-oxidation element – Si predestine this material for the microstructural degradation processes.

The microstructure of the tested materials is presented in Figs. 1-4. The static tensile results are collected in Table 2.

Table 2. Static tensile test results

Material	f_y [MPa]	f_u [MPa]	E [GPa]	A_5 [%]	Z [%]
Eiffel Bridge	292	342	193	8.1	11.6
Puddle iron 1863 [2]	287	360	191	15.3	33.9

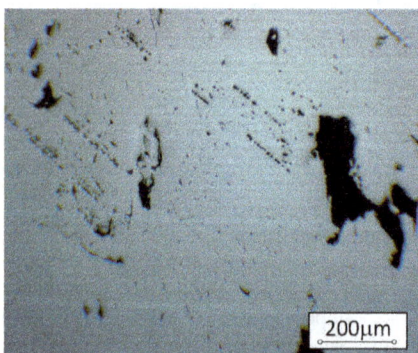

Fig. 1. Representative image of the material (Eiffel Bridge) structure in the non-etched state. Noticeable large non-metallic inclusions and slags, the size of the slags often exceeds mm, non-etched state [8].

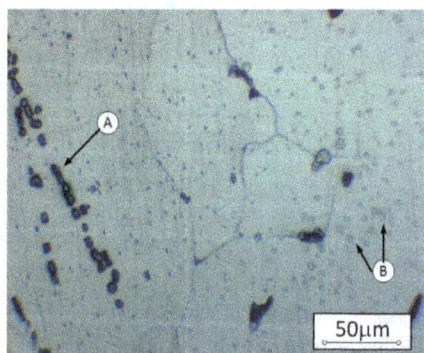

Fig. 2. Enlarged ferrite grains with non-metallic inclusion chains (A). Noticeable different grain size with brittle phase precipitations inside ferrite grains (B). Eiffel Bridge, etched 5%HNO₃ [8].

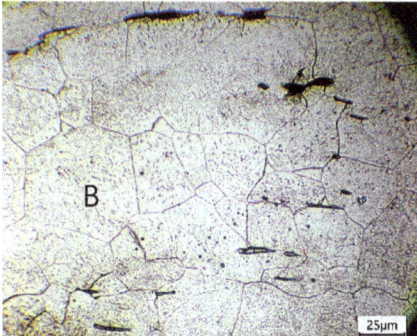

Fig. 3. Enlarged ferrite grains (1863 iron) with non-metallic inclusion chains and degradation precipitations (B), etched 5 %HNO$_3$ [6].

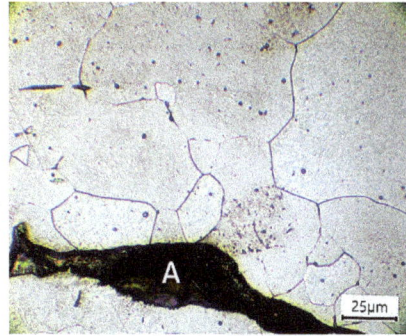

Fig. 4. Magnified ferrite grains with large non-metallic inclusions (A) in iron from viaduct (1863), etched 5%HNO$_3$ [6].

The microstructure of the investigated materials is shaped by numerous non-metallic inclusions (mainly silicates) and consist of ferrite grains of different size. The presence of the microstructural degradation processes is also confirmed by brittle phase precipitations occurring inside ferrite grains as well as on the grain boundaries. The microstructure of the tested material confirms the fact that the delivered for investigation material is an old puddle iron from the 19th-century – similar like in the previous research findings associated with old steel reported in [7].

Fatigue crack growth rate
In order to predict the remaining fatigue crack growth lifetime, it is essential to know the kinetics of fatigue crack growth. For this purpose, the fatigue crack growth rate was investigated. The used tensile machine, apparatus (e.g. clevises, grips etc.) and specimens (Fig. 5) were same as described in ASTM E647 [9]. The graphs of fatigue crack growth kinematic are presented in coordinates $log(da/dN) – log\Delta K$ (K_{max}), where, in logarithmic scale, the value of da/dN for specified crack length is marked on the ordinate and ΔK, K_{max} on the abscissa. The SIF for the CT specimen is specified using formula presented in ASTM E647 [9]:

$$K = \frac{\Delta F}{B\sqrt{W}} f\left(\frac{a}{W}\right), \tag{1}$$

$$f\left(\frac{a}{W}\right) = \frac{(2 + a/W)(0.886 + 4.64a/W - 13.32(a/W)^2 + 14.72(a/W)^3 - 5.6(a/W)^4)}{\sqrt{(1 - a/W)^3}} \tag{2}$$

where: a - crack length; B - specimen thickness; W - specimen width; ΔF - force amplitude.
During the examination were registered the following signals: force, displacements, crack opening displacement (COD). Amid applying of monotonically arising loading, the crack length size was determined by compliance procedures. The function of plane stress elastic compliance for CT specimens is described by the formula [9]:

$$\frac{a}{W} = C_0 + C_1 u_x + C_2 u_x^2 + C_3 u_x^3 + C_4 u_x^4 + C_5 u_x^5. \tag{3}$$

Experimental Mechanics of Solids   Materials Research Forum LLC
Materials Research Proceedings **12** (2019) 77-83   https://doi.org/10.21741/9781644900215-11

Coefficients C_0, C_1, C_2, C_3, C_4, C_5 are fully described by ASTM E647 [9] depending on measurement localisation of COD (Crack Opening Displacement). The u_x quantity is defined as:

$$u_x = \frac{1}{\sqrt{\frac{BEv}{F}} + 1}$$

(4)

where: B - specimen thickness; E - elastic modulus; v - COD; F - applied force; v/F - displacement versus force curve slope measured during the test.

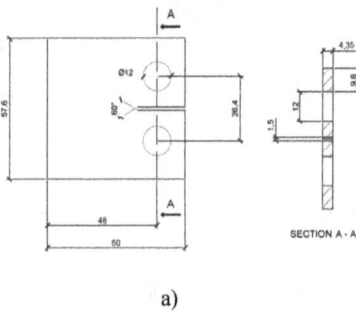

a)   b)

Fig. 5. a) CT specimen for mode I test (Eiffel Bridge) [8], b) machined CT specimens before test

Before the proper investigation, the fatigue pre-crack (mode I) was made preserving all condition of loading described in ASTM E647 [9]. During precracking the ΔK does not exceed 13-15 MPam$^{0.5}$ range. The main test started from the initial value of ΔK in order to avoid the plastic zone influence from the precracking procedure. Experiments were performed for two different mean stress levels characterised by the stress ratio $R = 0.05$ and $R = 0.5$ (for Eiffel Bridge) with sinusoidal loading waveform (frequency f=5 Hz). The crack length size was monitored using the phenomenon of plane stress elastic compliance (Eq. 3). The computer system was controlled by MTS FlexTest console and FCGR (Fatigue Crack Growth Rate) software integrated with MTS machine. Periodically, the crack length was controlled and adjusted using a stereoscopic microscope with a digital camera coupled with the tensile machine MTS 810. For the puddle iron from viaduct (1863) similar procedure (fully described in refs. [6,8]) was adopted. Due to material limitations, the size of the CT specimen was decreased. In order to evaluate the fatigue crack growth resistance for such material, tests were performed using two types of Compact Tension (CT) specimens with the main dimensions: W=38 mm, t=6 mm (thickness). The area of interest was the near-threshold region, of the FCGR curve and stable Paris regime using ΔK-control decreasing/increasing test (R=0.1, f=12Hz). Similarly as in the case of the Eiffel Bridge steel, a signal of the force, COD and displacement were registered.

During the experiment for chosen cycles at least two hysteresis loops were registered in order to assess the F-COD behaviour during test. According to the experimental procedure described in refs. [6,8], the crack closure point as well as corresponded F_{cl} and K_{cl} levels was estimated in order to evaluate the effective stress intensity range (ΔK_{eff}) based on the Elber concept [10].

Finally, the Kinetic Fatigue Fracture Diagrams were constructed for both materials – Fig. 6.

Experimental Mechanics of Solids
Materials Research Proceedings **12** (2019) 77-83

Materials Research Forum LLC
https://doi.org/10.21741/9781644900215-11

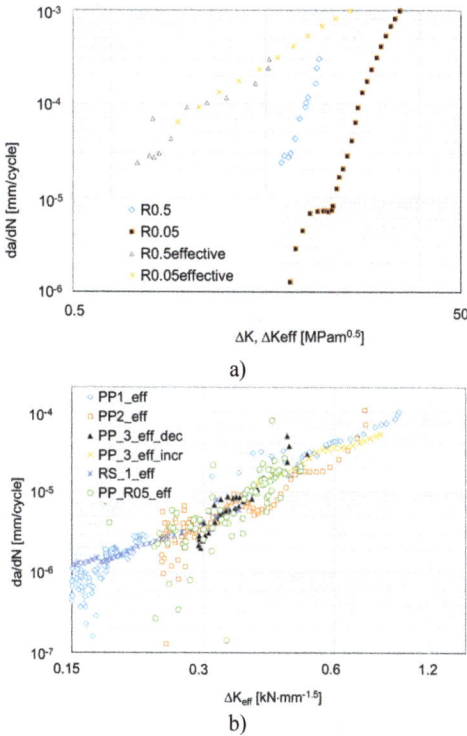

Fig. 6. Kinetic Fatigue Fracture Diagram for; a) Eiffel Bridge, b) puddle iron from viaduct
(1863) [6].

A noticeable is R-ratio effect in KFFD represented by shifted da/dN curves (Fig. 6a) for $R=0.5$ and $R=0.05$. In case of the ΔK_{eff}, the effect can be neglected in Paris regime. The same effect is observed for all R-ratios (0.05, 0.1 and 0.5) in fatigue crack growth results for puddle iron from viaduct. In the case of historic steel, the unification of the KFFD notation seems to be a key issue. One of the promising approaches is the energy approach proposed in refs. [11,12]. In the range of linear fracture mechanics a similar effect (R-ratio avoidance) was achieved as for effective ΔK using ΔS^* [11]. It is worth noting that during a cyclic load, J_{max} and ΔJ are the values that bind the local stress intensity in front of the crack front - J plays the same role as K in the elastic-plastic fracture mechanics. Therefore, it the proposed new crack driving force [11] is described as follows:

$$\Delta S^* = \sqrt{\Delta J^+ \cdot J_{max}}. \tag{5}$$

For energy fatigue crack growth description (puddle iron from Eiffel bridge) only elastic part of ΔJ-integral range was analysed using well-known relationship from linear-elastic fracture mechanics (with assumed plane stress conditions):

$$\Delta J_e = \frac{\Delta K^2}{E}. \tag{6}$$

The new energy-based kinetic fatigue fracture diagrams are presented in Fig. 7a (based only on cyclic J-integral parameter) and Fig. 7b (based on new ΔS^* parameter). As it is noticeable the new ΔS^* strain energy parameter describes kinetics of fatigue crack growth synonymously and independently from R-ratio.

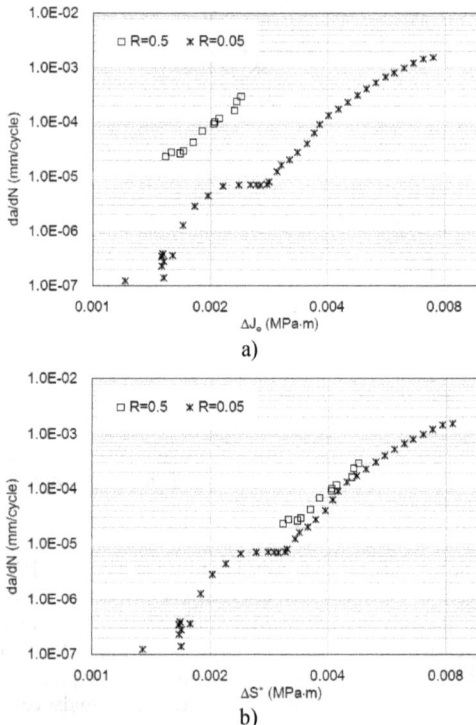

Fig. 7. Kinetic Fatigue Fracture Diagram (KFFD) for Eiffel Bridge steel based on; a) elastic ΔJ_e parameter, b) ΔS^* strain energy density parameter.

Conclusions

Based on the performed experimental tests, some conclusions can be formulated. The Paris' exponent m for old puddle iron seems to be significantly higher than for modern bridge constructional steel. The shifting of the da/dN-$\Delta K_{applied}$ curves was observed for different R-

Experimental Mechanics of Solids
Materials Research Proceedings 12 (2019) 77-83

Materials Research Forum LLC
https://doi.org/10.21741/9781644900215-11

ratios. Furthermore, the observed fatigue crack closure phenomenon strongly influence the kinetics of fatigue crack growth (via decrease Paris' m exponent and consolidate data from different R-ratios into one curve). The proposed, new energy parameter ΔS^* describes the kinetics of fatigue crack growth in puddle iron effectively without mean stress effect (R-ratio).

References

[1] G. Lesiuk, B. Rymsza, J. Rabiega, J.A.F.O. Correia, A.M.P. De Jesus, R. Calçada. Influence of loading direction on the static and fatigue fracture properties of the long term operated metallic materials. Eng Fail Anal 2019;96:409-425. https://doi.org/10.1016/j.engfailanal.2018.11.007

[2] B. Pedrosa, J.A.F.O. Correia, et al. Fatigue resistance curves for single and double shear riveted joints from old Portuguese metallic bridges. Eng Fail Anal 2019; 96:255-273. https://doi.org/10.1016/j.engfailanal.2018.10.009

[3] G. Lesiuk, J.A.F.O. Correia, et al. Fatigue crack growth rate of the long term operated puddle Iron from the Eiffel bridge. Metals 2019; 9(1), Article No. 53. https://doi.org/10.3390/met9010053

[4] G. Lesiuk. Application of a new, energy-based ΔS^* crack driving force for fatigue crack growth rate description. Mater 2019;12(3), Article No. 518. https://doi.org/10.3390/ma12030518

[5] ASTM E407-07(2015)e1 Standard Practice for Microetching Metals and Alloys, ASTM International, West Conshohocken, PA, 2015, https://doi.org/10.1520/E0407-07R15E01

[6] G. Lesiuk, M. Szata, J. A. Correia, A. M. P. De Jesus, F. Berto, (2017). Kinetics of fatigue crack growth and crack closure effect in long term operating steel manufactured at the turn of the 19th and 20th centuries. Engineering Fracture Mechanics, 185, 160-174. https://doi.org/10.1016/j.engfracmech.2017.04.044

[7] G. Lesiuk, M. Szata , M. Bocian (2015). The mechanical properties and the microstructural degradation effect in an old low carbon steels after 100-years operating time. *Archives of Civil and Mechanical Engineering*, *15*(4), 786-797. https://doi.org/10.1016/j.acme.2015.06.004

[8] G. Lesiuk, J. Correia, M. Smolnicki, A. De Jesus, M. Duda, P. Montenegro, R. Calcada, (2019). *Fatigue Crack Growth Rate of the Long Term Operated Puddle Iron from the Eiffel Bridge*. Metals, 9(1), 53. https://doi.org/10.3390/met9010053

[9] ASTM International, 2015. ASTM E647 - 15 Standard Test Method for Measurement of Fatigue Crack Growth Rates. In United States: ASTM International, p. 43. Available at: http://www.astm.org/Standards/E647. https://doi.org/10.1520/jai13180

[10] W. Elber 1970, Fatigue crack closure under cyclic tension, Engineering Fracture Mechanics, 2, pp. 37-45, 1970. https://doi.org/10.1016/0013-7944(70)90028-7

[11] G. Lesiuk (2019). Mixed mode (I+II, I+III) fatigue crack growth rate description in P355NL1 and 18G2A steel using new energy parameter based on J-integral approach, Engineering Failure Analysis, 93, 263-272. https://doi.org/10.1016/j.engfailanal.2019.02.019

[12] G. Lesiuk, M. Szata, D. Rozumek, Z. Marciniak, J. Correia, A. De Jesus, (2018). Energy response of S355 and 41Cr4 steel during fatigue crack growth process. The Journal of Strain Analysis for Engineering Design, 53(8), 663-675. https://doi.org/10.1177/0309324718798234

Experimental Mechanics of Solids
Materials Research Proceedings **12** (2019) 84-89

Materials Research Forum LLC
https://doi.org/10.21741/9781644900215-12

Modeling of Neck Effect in Cylindrical Shell

Adam Piwowarczyk[1,a], Artur Ganczarski[2,b*]

[1]Institute of Applied Informatics, Cracow University of Technology, Al. Jana Pawła II 37, 30-860 Kraków, Poland

[2] Institute of Applied Mechanics, Cracow University of Technology, Al. Jana Pawła II 37, 30-860 Kraków, Poland

[a]adam.piwowarczyk@mech.pk.edu.pl, [b]artur.ganczarski@pk.edu.pl

Keywords: Neck Effect, Cylindrical Shell, Low Cycle Fatigue

Abstract. This work presents two models of a neck effect appearing in a cylindrical shell. In the first case, an elastic deformation is described by the conventional shell equation in which axial force appears as an external loading. In the second case, an elastic-plastic deformation including damage effect is described by the kinetic theory of damage evolution implemented to FEM code. Results obtained from both models confirm existence of a distinct displacement, stress and damage localization in the cylindrical shell free from any geometrical or material imperfections.

Introduction

A final stage of the uniaxial tension test is accompanied by characteristic local contraction of a specimen called neck effect. Aforementioned phenomenon is well visible in elongation-force or strain-nominal stress diagram since it corresponds to the beginning of unstable deformation, usually associated with reaching of the ultimate strength. During the necking prior to fracture, originally uniaxial and homogeneous stress state in a specimen becomes inhomogeneous and tri-axial. In case of circular specimen analysis of 3D stress state was originally presented by Bridgman [1] and independently by Davidenkov and Spiridonova [3]. In early sixties Cowper and Onat published the paper [2] on neck effect accompanying plane state of strain. In case of a thin-walled tubular specimen the stress state was analyzed by Malinin and Rżysko [6] by use of Laplace's equation. Original numerical analysis of necking process in a bar made of material subjected to rigid-plastic hardening was presented by Dietrich [11]. The assumption of specific format of local displacement rate field was used next to determine a slip system and a range of plastic zone preceding the neck formation.

Generally, the neck problem is still treated as complicated and not sufficiently recognized. The main reason is inhomogeneity of plastic properties, generally exhibiting random distribution, which causes that location of neck is not known in advance. As a consequence, in majority of approaches the neck is treated as a notch of a priori known profile and location.

General formulation of cylindrical shell

Elastic range

A thin-walled cylindrical shell of variable thickness is considered as a prototype of the tubular specimen made of aluminum alloy Al-2024 shown in Fig. 1.

Experimental Mechanics of Solids
Materials Research Proceedings 12 (2019) 84-89

Materials Research Forum LLC
https://doi.org/10.21741/9781644900215-12

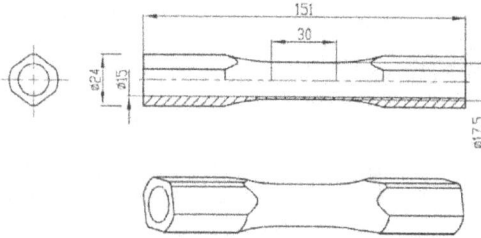

Fig. 1. Geometry of tubular specimen made of aluminum alloy Al-2024.

Conventional differential equation of the cylindrical shell subjected to axial force is as follows

$$\frac{d^2}{dx^2}\left[\frac{Eh^3(x)}{12(1-v^2)}\frac{d^2w}{dx^2}\right] + \frac{Eh(x)}{R^2}W = -\frac{vn}{R},\qquad(1)$$

where the right hand side is of special format. Namely, although the problem under consideration deals with the shell loaded by axial force n exclusively, Poisson's effect $-vn/R$ formally corresponds to external pressure. In case of clamped edge of shell, what refers to the tubular specimen having shoulders filled in by pivots, aforementioned term is directly responsible for displacement localization.

Elastic-plastic range including damage

Elastic-plastic range including damage effect of the cylindrical shell is described by use of kinetic theory of damage evolution invented by Lemaitre and Chaboche [5]. Key point of this formalism is dissipation potential

$$F = \left[\frac{3}{2}(\tilde{s}-X'):(\tilde{s}-X')\right]^{1/2} - R - \sigma_y + \frac{3}{4X_\infty}X':X' + \frac{Y^2}{2S(1-Dh)}H(p-p^D),\qquad(2)$$

where $\tilde{s}$ stands for the deviator of effective stress, which takes format

$$\tilde{s} = \frac{\sigma - \frac{1}{3}\mathrm{Tr}(\sigma)1}{1-Dh}.\qquad(3)$$

The last part of the dissipation potential, called the damage potential, is a function of cumulative plastic strain p activated, when damage threshold p^D is attained. The thermodynamic force conjugated to damage Y is equal to the amount of the elastic energy density

$$Y = \frac{1}{2}E^{-1}:\tilde{\sigma}:\tilde{\sigma}.\qquad(4)$$

Then applying the formalism of associated plasticity one can find increments of following variables: plastic strain $d\varepsilon^p$, cumulative plastic strain dp, strain conjugated to isotropic dr and kinematic hardening $d\alpha$, damage dD

$$d\varepsilon^{\mathrm{p}} = \frac{\partial F}{\partial \sigma} d\lambda = \frac{3}{2} \frac{\tilde{s}-X'}{[\frac{3}{2}(\tilde{s}-X'):(\tilde{s}-X')]^{1/2}} \frac{d\lambda}{1-Dh}, \qquad dp = (\frac{2}{3} d\varepsilon^{\mathrm{p}}: d\varepsilon^{\mathrm{p}})^{1/2} = \frac{d\lambda}{1-Dh},$$

$$dr = -\frac{\partial F}{\partial R} d\lambda = (1-Dh)dp, \qquad d\alpha = -\frac{\partial F}{\partial X'} d\lambda = d\varepsilon^{\mathrm{p}}(1-Dh) - \frac{3}{2X_{\infty}} X' d\lambda, \qquad (5)$$

$$dD = \frac{\partial F}{\partial Y} d\lambda = \frac{Y}{S} H(p-p^{D})dp.$$

In the next step, the inner variables r, α, and Y appearing in (5) are expressed by subsequent thermodynamic forces R, X and D, calculated form the potential of free energy

$$\rho\psi = \frac{1}{2} E: \varepsilon^{\mathrm{e}}: \varepsilon^{\mathrm{e}}(1-Dh) + R_{\infty}[r + \frac{1}{b}\exp(-br)] + \frac{\gamma X_{\infty}}{3} \alpha: \alpha. \qquad (6)$$

Namely, taking appropriate derivatives of (6) with respect to ε^{e}, r and α, it is possible to convert kinematic inner variables of equations (5) to their stress like equivalents

$$\sigma = \rho\frac{\partial\psi}{\partial\varepsilon} = E: \varepsilon^{\mathrm{e}}(1-Dh), \quad R = \rho\frac{\partial\psi}{\partial r} = R_{\infty}[1-\exp(-br)],$$

$$X = \rho\frac{\partial\psi}{\partial\alpha} = \frac{2}{3}\gamma X_{\infty}\alpha, \qquad Y = -\rho\frac{\partial\psi}{\partial D} = \frac{1}{2} E: \varepsilon^{\mathrm{e}}: \varepsilon^{\mathrm{e}}h, \qquad (7)$$

achieving final format of constitutive equations for plastic range

$$\sigma = E: (\varepsilon - \varepsilon^{\mathrm{p}})(1-Dh), \qquad dR = b(R_{\infty}-R)d\lambda,$$

$$dX' = \gamma[\frac{2}{3}X_{\infty}d\varepsilon^{\mathrm{e}}(1-Dh) - X'd\lambda], \quad dD = \frac{E^{-1}:\sigma:\sigma}{2S(1-Dh)^2} H(p-p^{D})dp. \qquad (8)$$

The magnitude of the plastic multiplier $d\lambda$ is calculated from the compatibility condition

$$\frac{\partial F}{\partial s}: ds + \frac{\partial F}{\partial X'}: dX' + \frac{\partial F}{\partial R}: dR = 0, \qquad (9)$$

and it is equal to

$$d\lambda = \frac{\frac{3}{2}(\tilde{s}-X')d\sigma'}{(1-Dh)[\frac{3}{2}(\tilde{s}-X'):(\tilde{s}-X')]^{1/2}[\gamma X_{\infty}+b(R_{\infty}-R)-\frac{3}{2}(\tilde{s}-X')(\frac{\tilde{s}}{1-Dh}\frac{\partial F}{\partial Y}+\gamma X')]}. \qquad (10)$$

The damage deactivation parameter h is defined by formula

$$h(\sigma) = h_c + (1-h_c)\frac{\chi(\sigma)-\chi(\sigma_{\mathrm{e}})}{\chi(\sigma_{\mathrm{b}})-\chi(\sigma_{\mathrm{e}})}, \qquad (11)$$

where Hayhurst's function $\chi(\sigma)$ is used to combine uniaxial damage D and tridimensional state of stress σ (see Cegielski [8], Ganczarski and Cegielski [9]).

Results
In case of the cylindrical shell working in elastic range, the fourth order differential equation (1) is numerically integrated from $x_1 = 0$ mm to $x_2 = 75$ mm by a use of the finite difference routine as well as the fourth-order Runge-Kutta technique (see Press et al. [10]), which require application of following boundary conditions

$$w(x_1) = 0, \quad w'(x_1) = \frac{dw}{dx}|_{x_1} = 0, \quad w''(x_2) = \frac{d^2w}{dx^2}|_{x_2} = 0, \quad w'''(x_2) = \frac{d^3w}{dx^3}|_{x_2} = 0. \quad (12)$$

Experimental Mechanics of Solids   Materials Research Forum LLC
Materials Research Proceedings **12** (2019) 84-89   https://doi.org/10.21741/9781644900215-12

Distributions of displacement function $w(x)$ and its second differential $w''(x)$, directly corresponding to the bending moment $m_x(x) = \frac{Eh^3(x)}{12(1-\nu^2)}w''(x)$, exhibit local extrema located in relatively narrow range $x \in (55,60)$, referring to the gauge length adjacent to transition region of conical profile (see Fig. 2).

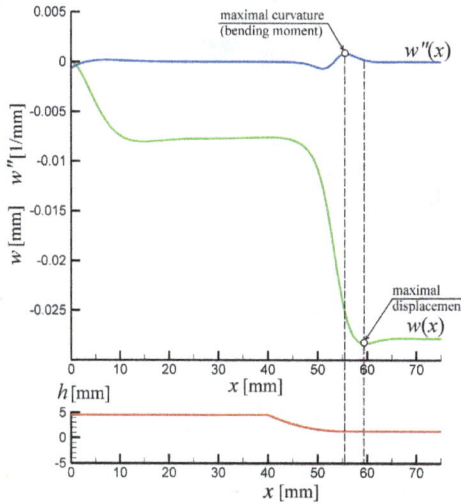

Fig. 2. Distributions of displacement function $w(x)$ and its second differential $w''(x)$ compared to profile of cylindrical shell.

In case of the cylindrical shell working in elastic-plastic range, system of equations (8-11) is implemented to the Finite Element Method (FEM) code for quadrilateral axisymmetric element (see Owen and Hinton [7], Ganczarski and Skrzypek [4]) .
Complete material data of the aluminum alloy Al-2024 is presented in Tab. 1.

Tab. 1. Material data of the aluminum alloy Al-2024 (see Ganczarski and Cegielski [9]).

E[GPa]	σ_y[MPa]	b	γ	ν	p^D	S[GPa]	R_∞[MPa]	X[MPa]	ρ[kg/m^3]
70	230	0.1	4.0	0.3	0.248	3.5	120	60	$2.7 \cdot 10^6$

Two types of displacement-control low cycle fatigue processes are analyzed: a pendulum like tension-compression $\Delta l = \pm 0.75$ cm and a pulsating tension $0 \le \Delta l \le 0.75$ cm. Similarly to the elastic range, first nucleation of plasticity and corresponding damage takes place in the gauge length adjacent to transition region of conical profile. However, combined effects of isotropic and kinematic plastic hardening cause motion of the damage front, which finally localizes in the middle of gauge length (see Fig. 3).

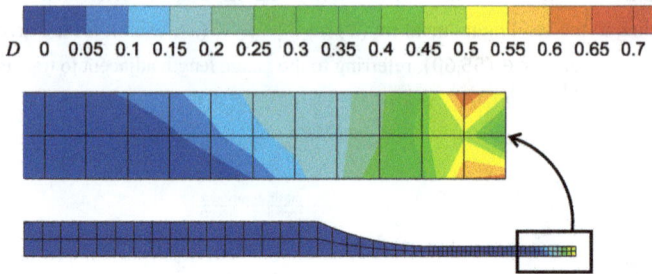

Fig. 3. Distribution of damage D in the whole mesh and localization region.

Displacement field corresponding to the initiation of first micro-crack exhibits clear localization (see Fig. 4).

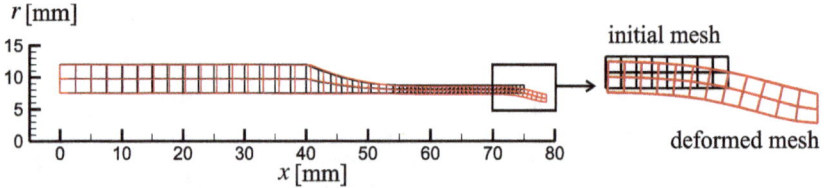

Fig. 4. Mesh: initial (black) and deformed (red) with localization region.

In case of the pendulum-like tension-compression cycle (see Fig. 5a) displacement-force diagram performs regular hysteresis loops, which exhibit stronger effect of a force amplitude decrease

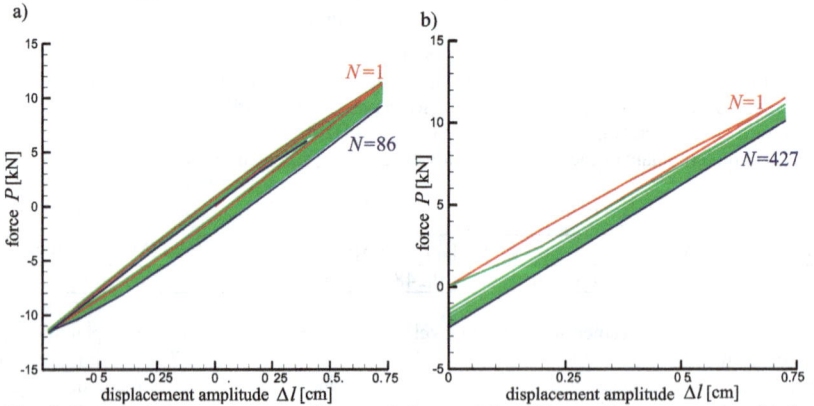

Fig. 5. Hysteresis loops corresponding to: a) the pendulum-like tension-compression, b) the pulsation cycles.

Experimental Mechanics of Solids
Materials Research Proceedings 12 (2019) 84-89

Materials Research Forum LLC
https://doi.org/10.21741/9781644900215-12

under tension than compression. This is the consequence of damage deactivation effect and total number of cycles to failure is equal to $N=86$. In case of the pulsation tension test (see Fig. 5b) subsequent hysteresis loops are more oblate and exhibit symmetric drop of the force amplitude, leading to characteristic drift of the diagram towards negative magnitudes of the force P. Since now the amplitude of displacement Δl is equal to one half of the previous one, corresponding total number of cycles to failure is bigger and equal to $N=427$.

Summary
Results obtained in numerical modeling of the cylindrical shell made of Al-2024 fully confirm that neither geometrical nor material imperfections are necessary to initiate the neck effect. Both examples presented in this work, comprising the cylindrical shell working: in elastic range or in elastic-plastic range including damage, are solved by the use of three independent numerical routines: the finite difference scheme, a direct integration with the fourth-order Runge-Kutta technique and the FEM code. The first example shows location of the neck in very narrow zone between the gauge length and the adjacent conical profile. Nevertheless, combined effects of plastic hardening and the damage, cause the motion of neck which finally localizes in the middle of gauge length.

Acknowledgments
The authors gratefully acknowledge the support rendered by the Institute of Applied Informatics and the Institute of Applied Mechanics, Cracow University of Technology.

References
[1] P.W. Bridgman, The stress distribution at the neck of a tension specimen, Trans. ASME, 32 (1944) 533-574.

[2] C.R. Cowper, E.T. Onat, The initiation of necking and buckling in plane plastic flow, Proc. 4th U.S. Nat. Congr. Appl. Mech. (1962) 1023-1029.

[3] N.N. Davidenkov, N.E. Spiridonova, Analysis of stress state in the neck of tensile specimen, Zavodska Laboratoria, Vol. XI, No. 6 (1945) (in Russian).

[4] A. Ganczarski, J. Skrzypek, Plasticity of engineering materials, Cracow University of Technology Publ. House, Kraków, 2007 (in Polish).

[5] J. Lemaite, J.-L. Chaboche, Méchanique des matériaux solides, Bordas, Paris, 1985.

[6] N.N. Malinin, J. Rżysko, Mechanics of materials, PWN, Warsaw, 1981 (in Polish).

[7] D.R.J. Owen, E. Hinton, Finite elements in plasticity: Theory and Practice, Pineridge Press Ltd., Swansea, 1980.

[8] M. Cegielski, Effect of continuous damage deactivation in CDM, PhD Thesis of Cracow University of Technology, 2011 (in Polish).

[9] A. Ganczarski, M. Cegielski, Continuous damage deactivation in modeling of cycle fatigue of engineering materials, Procedia Eng. 2(1) (2010) 1057-1066. https://doi.org/10.1016/j.proeng.2010.03.114

[10] W. Press W., S. Teukolsky, W. Vettering, B. Flannery, Numerical recipes in Fortran, Cambridge Press, Cambridge, 1993.

[11] L. Dietrich, Consideration of material hardening in analysis of combined plastic flow processes, Issues of IPPT PAN, No 53, 1977 (in Polish).

Experimental Mechanics of Solids
Materials Research Proceedings **12** (2019) 90-95

Materials Research Forum LLC
https://doi.org/10.21741/9781644900215-13

Selected Aspects of Stand Tests for Prototype Floating Bridge Joints

Wieslaw Krason[1,a *], Pawel Bogusz[2,b], Arkadiusz Poplawski[3,c] and Michal Stankiewicz[3,c]

[1]Military University of Technology, Gen. Witolda Urbanowicza Street 2, 00-908 Warsaw, Poland

[a]wieslaw.krason@wat.edu.pl, [b]pawel.bogusz@wat.edu.pl, [c]poplawski.arkadiusz@wat.edu.pl, [d]michal.stankiewicz@wat.edu.pl

Keywords: Floating Cassette, Rod Joint, Strength Tests, Three-Point Bending Test, Force-Displacement Curve, Joint Deformation, DIC Method

Abstract. The subject of the research is rod joints mounted in a prototype version of a floating module in the form of a metal cassette with adjustable buoyancy. Such joints consist of a movable mandrel and sleeves with a closed cross-section. Experimental stand for a three-point bending test is built on the base of the testing machine. The force-displacement curves are determined for various load cases and the joint working positions. Deformations and displacements of the connector components are also recorded using high-speed cameras. The influence of a blind screw blocking the movement of the mandrel in the connector sleeve on operation in various construction configurations is determined based on the results.

Introduction

Floating bridges belong to the group of special bridges with high mobility. They are most often used by the army to build a temporary water crossing. An important role in the operation of floating bridges is performed by side connectors. They enable fast connection of the same floating segments in a ribbon bridge. Side joints should provide high durability as well as proper operating parameters of the assembled bridge structure. The prototype floating cassettes with adjustable buoyancy [1, 2] are built at the Military University of Technology. Construction works, numerical and experimental tests of the floating crossing system subassemblies are continued [2, 3].

The main connector of the prototype floating cassette shown in Fig. 1a is the subject of the research presented in the paper. The strength joints testing under operating conditions of a complete bridge is difficult (these joints are usually submerged in the water), costly and technically impossible to perform due to the safety of the crew and equipment. Therefore, the research methodology and selected test results of the prototype main connector separated from the bridge cassette - Figure 1b are discussed in the paper. Such joint [4] consist of a movable mandrel and sleeves with a closed cross-section. The connector mandrel locked in the sleeve of one cassette is inserted into the sleeve of the next cassette attached to the ribbon. Such joints can operate as main connectors (mounted vertically on the sides of the cassette) and additional connectors (operating horizontally in the roadway of the cassette) - Fig. 1a. In both of the above cases, these joints work between adjacent cassettes mainly for bending. Therefore, a three-point bending stand of a separated joint on the basis of a testing machine with a sufficiently wide working space is built. An assembly drawing and a separate joint arranged horizontally on the stand are shown in Fig. 1 b, c.

Experimental Mechanics of Solids   Materials Research Forum LLC
Materials Research Proceedings **12** (2019) 90-95   https://doi.org/10.21741/9781644900215-13

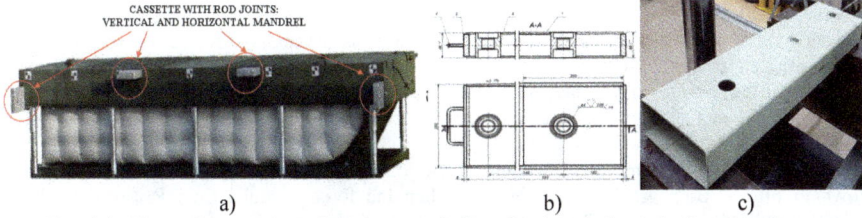

Fig. 1a). View of a cassette with rod joints on the sides (vertical mandrel) and in the cassette
roadway (horizontal mandrel), 1b) joint drawing, c) separated joint for stand tests

During the laboratory tests, the force-displacement' curves are determined for the joint in
various settings, supports and structural configurations (with or without a mandrel locking by a
set blind screw). The maximum displacement values of the joint elements at the stand during the
load tests are additionally recorded using a high-speed camera. Displacements of the selected
points of the bent joint are also determined along its length and height based on the results
obtained by digital image correlation (DIC) methods. The comparison of displacement results
will allow for evaluation of flexibility of the subassembly components of the connector stand
during testing. The influence of the blind screw locking the movement of the mandrel in the
connector sleeve on the operation of these components in various structure configurations of the
connector is determined on the basis of the results

Stand for laboratory tests of the joint
Strength testing stands for the bridge joint are built. A strength machine is used for this purpose.
A separated joint for testing has been attached to the machine base. The movable machine punch
generated the joint bending force. Fig. 2 shows a diagram of the stands built for three-point
bending tests of a joint.

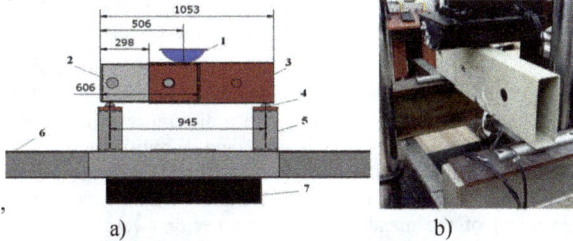

Fig. 2. Structure of the station for experimental investigations of the joint: a) diagram of the
three-point bending stand, b) separate rod joint prepared for the three-point bending test

A joint separated for tests is performed based on the design documentation of the cassette.
The basic components and the structure of the three-point bending stand for testing of the
separated joint on the strength machine are shown in Fig. 2a. The diagram shows the basic
components of the three-point testing stand: 1-punch of the testing machine, 2 - mandrel of the
tested joint, 3 - sleeve joint, 4 - half-shaft, 5 - moveable base, 6 - I section beam, 7 - base of the
testing machine. A stand ready for testing with the joint set in a vertical position is shown in
Fig. 2b. An additional equipment (cameras, lighting system and others) used to determine
displacements and deformation of subassemblies is shown in Fig. 4a.

Experimental Mechanics of Solids   Materials Research Forum LLC
Materials Research Proceedings **12** (2019) 90-95   https://doi.org/10.21741/9781644900215-13

Research results

Laboratory tests of three-point bending are carried out using the above discussed stand. The rod joint positioned in various settings is bent by a vertical force with different values generated by the punch of the strength testing machine. The 'Force-Displacement P-V' curves are determined for joints in various settings, supports and structure configurations (with or without a mandrel locking by a set blind screw).

Selected results of a three-point bending of the joint placed vertically on the test stand are shown in Fig. 3. Both graphs were obtained when the force changed from $P_{min} = 0$ to the maximum value $P_{max} = 100kN$. The wider hysteresis loop, shown in Figure 3a, refers to the test case of the joint with the blind screw locking the mandrel position in the connector sleeve. The deformation of the joint after the unloading process is noticeable. The value of displacements recorded after unloading to $P_{min} = 0$ is up to $V = 1mm$. It may result from blocking the mandrel in the sleeve after resetting the nominal clearance due to the limited rotation of the mandrel against the locking blind screw- Fig. 4b.

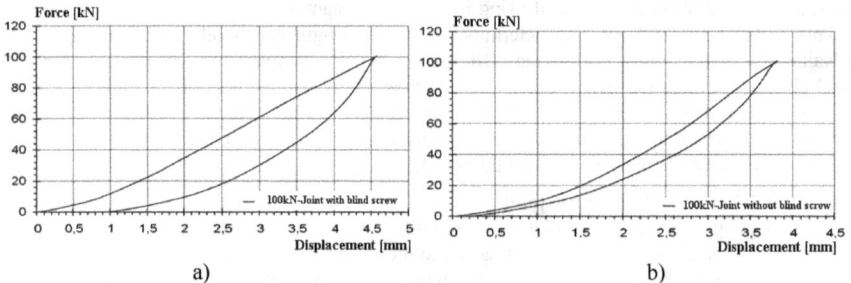

a)   b)

Fig. 3 Graphs of load changes as a function of joint deflection in a three-point bending test with force of 100kN: a) with a blind screw blocking the position of the mandrel in the connector sleeve, b) without a set blind screw

The tested joint consists of components that can move relative to each other in a range limited by the initial clearances. These gaps occur due to a difference in nominal dimensions of the mandrel and the connector sleeve. They enable and simplify the connection of the joint components by inserting the mandrel into the sleeve - Fig. 1b, c. A detailed analysis of the joint components displacements towards the horizontal and vertical axis is necessary due to the independent movement of the mandrel and the connector sleeve. Therefore, changes in the displacement values of selected joint elements at the stand during the loading tests are additionally recorded using a high-speed camera - Fig. 4a. Five markers named Point 1-5 (Fig. 4b) are glued on the lateral surface of the mandrel and the sleeve of the joint set vertically. They are used to register the displacement components of the joint subassemblies with cameras, using digital image correlation (DIC) method. The graphs describing changes in the components of the joint displacements during load tests are obtained on the basis of the data recorded with the camera and processed using the Tema software. Selected results of the changes in vertical and horizontal displacements during the three-point bending test of the joint with the maximum force $P_{max} = 100kN$ are shown in Fig. 5-8.

The diagrams of the vertical displacements changes recorded by the camera for the subassemblies of the joints with the blind screw locking the position of the mandrel in the sleeve are shown in Fig. 5. Fig. 6 presents the analogous results obtained in the joint tests without the

Experimental Mechanics of Solids   Materials Research Forum LLC
Materials Research Proceedings **12** (2019) 90-95   https://doi.org/10.21741/9781644900215-13

set blind screw. The graphs of the horizontal displacements changes recorded by the camera for the joint subassemblies respectively with the blind screw locking the position of the mandrel in the sleeve, and the diagrams of the horizontal displacement changes registered in the joint components during tests without the set blind screw are presented in Fig. 7 and Fig. 8, respectively.

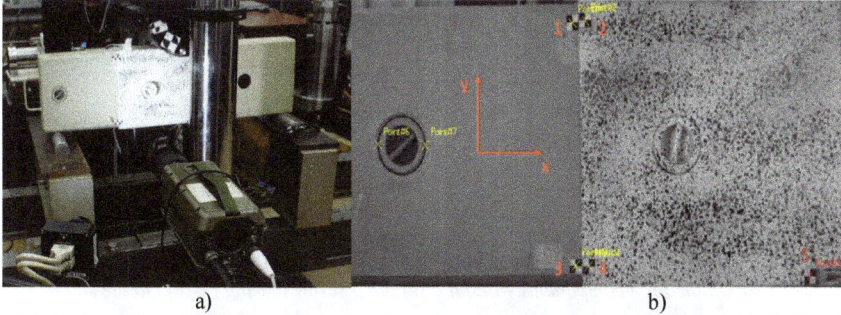

a)   b)

Fig. 4. Stand for registering independent displacements of the joint components with the blind screw; a) an equipment of the test stand with cameras, b) the tested joint with the set blind screw locking the mandrel against the sleeve and markers described with red numbers in the range of 1 - 5

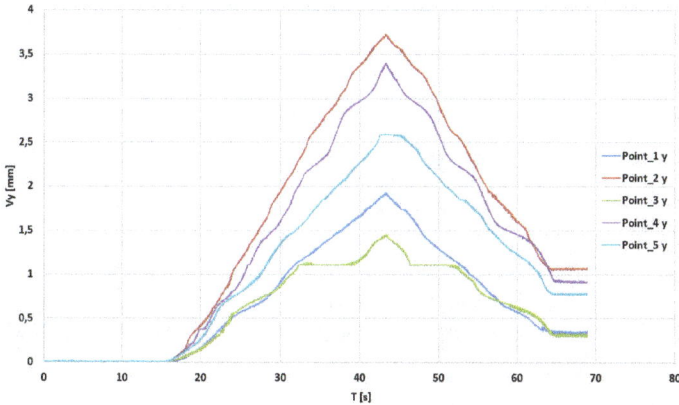

Fig. 5. Graphs of the vertical displacements changes registered for various measuring points (Fig. 4b) of the joint components with the blind screw locking the mandrel with respect to the sleeves

The maximum vertical displacements are recorded in point 2 (Fig. 4b) on the side wall of the joint sleeve in the test variant with the blind screw locking the mandrel. They are $V_Y = 3.7$mm. The maximum horizontal displacement equal to $V_X = 0.7$mm is recorded in point 5 (Fig. 4b), also on the side wall of the joint sleeve during the test variant with the set blind screw.

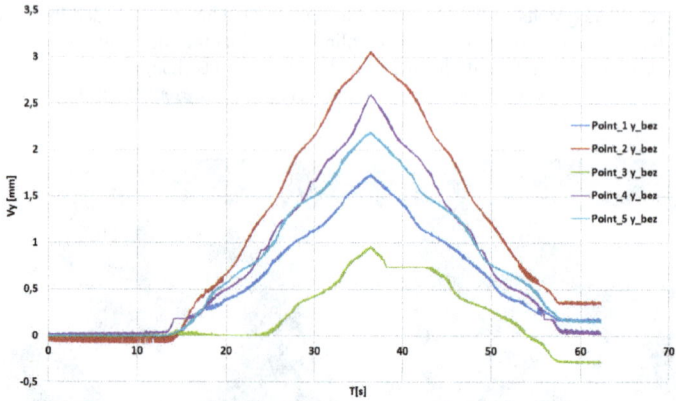

*Fig. 6. Graphs of the vertical displacements changes registered for various measuring points
(Fig. 4b) in the joint components without a set blind screw locking respect to the sleeve*

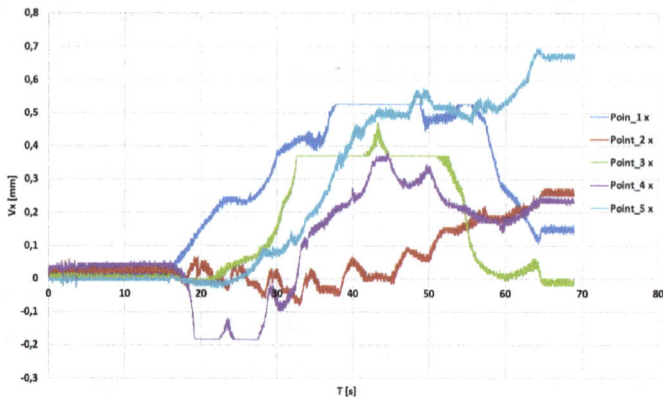

*Fig. 7. Graphs of the horizontal displacements changes registered for various measuring points
(Fig. 4b) in the joint components with a set blind screw locking respect to the sleeve*

Summary and Conclusions

The laboratory stand built on the basis of the Satec1200 testing machine is used for a three-point
bending test of the separated joint. The load tests of the joint separated from the cassette are
developed with the vertical settings of the mandrel (as in the main rod joint of the floating
cassette [3, 4]) and various structure configurations of the joint subassemblies. They correspond
to the use of a set blind screw between the mandrel and the joint sleeve and the free insertion of
the mandrel into a hole of the sleeve without the blind screw locking a mandrel

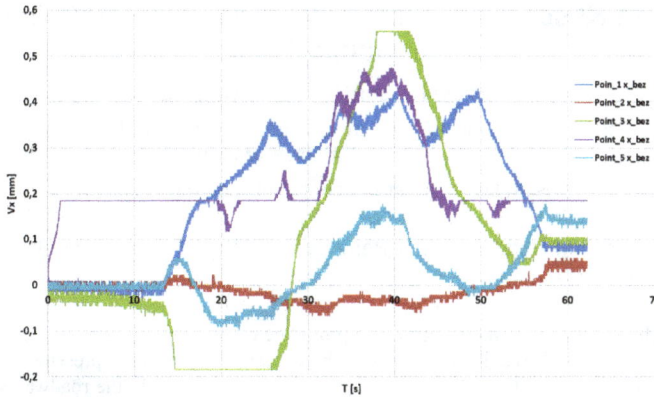

Fig. 8. Graphs of the horizontal displacements changes recorded in various measuring points (Fig. 4b) of the joint components without the blind screw locking the mandrel with respect to the sleeves

Displacements of the joint components are recorded using the control system of the strength machine ('Force-Displacement' P-V graphs) and with an optical system during the load tests. It is found that larger hysteresis loops appear in P-V curves in the case of the data registered in the joint test variants with the blind screw locking the position of the mandrel in the sleeve.

It is possible to independently track the movement of the joint components along the vertical axis and the horizontal plane of the image with the use of high-speed cameras. Based on the test results, it is also found that the lack of the blind screw does not cause a slide of the mandrel from the joint sleeve during the three-point bending test.

The result of measurements indicates a significant impact of the blind screw locking the mandrel in the joint sleeve on a mechanism of the joint components interaction (mainly the important role of the right size clearances between the mandrel and the sleeve).

Acknowledgement

The works were carried out as part of the research project PBS 23-937, Military University of Technology, 2016-2018.

References

[1] Military University of Technology, European Patent, No. 2251255, A sectional pontoon bridge. (2013)

[2] W. Krason, J. Malachowski, Field test and numerical studies of the scissors-AVLB type bridge, Bulletin of The Polish Academy of Sciences, Technical Sciences, Vol. 62, No. 1 (2014) 103-112. https://doi.org/10.2478/bpasts-2014-0012

[3] W. Krason, P. Slawek, Design and pre-testing of a mobile modular floating platform with adjustable displacement, Mechanik nr 11(2017) 1075-1080. https://doi.org/10.17814/mechanik.2017.11.185

[4] Military University of Technology, Patent Office of the RP, PAT.223689, Set of mechanical locks to connect the floating bridge cassettes and the cassette opening mechanism. (2016)

Experimental Mechanics of Solids   Materials Research Forum LLC
Materials Research Proceedings **12** (2019) 96-103   https://doi.org/10.21741/9781644900215-14

Strain Research of Floating Bridge Side Joints in Lab Loading Tests

Wiesław Krasoń[1,a*], Paweł Bogusz [1,b]

[1]Faculty of Mechanical Engineering, Military University of Technology,
00-908 Warsaw, Witolda Urbanowicza Street 2, Poland

[a]wieslaw.krason@wat.edu.pl, [b]pawel.bogusz@wat.edu.pl

Keywords: Floating Bridge, 3-Point Bending Test, Digital Image Correlation Method, Experimental Mechanics

Abstract. The separated main connector of a prototype cassettes floating bridge, composed of a mandrel and an outer sleeve, is the subject of the experimental research presented in the work. Operating conditions, similar to those found in side connectors and in the roadway surface of a real joint prototype, are mapped during the lab tests. Determination of strain maps of the prototype joint sleeve walls is the purpose of these tests. Strength tests of the three-point bending of the separated bridge side connector were preceded by a preliminary strength analysis with the use of analytical models and FEM analysis. The range of loads used is selected in such a manner that the maximum stress occurring in the joint sleeve walls should not exceed the yield strength of the material. The electroresistance gauge method is supplemented with the optical image correlation methods for deformation measurements. With a continuous strain distribution maps are developed in this manner. The strain maps enabled a precise identification of the most stressed areas in the joint sleeve wall during the bending tests, as well as a strength assessment of the structural solution of the blind screw fixing mandrel in the connector sleeve.

Introduction

Crossing bride constructions of various types are of great importance for the modern army mobility. They are successively developed, researched and implemented [1, 2, 3]. Modern armies are equipped with pontoon floating bridges [1, 4]. These structures are applicable both to the needs of the army and civilians. They enable organizing temporary crossings on rivers and various water reservoirs. Bridges of this type are used, among others, in places where fixed bridges are excluded from repair, due to, for example, operational damage. To replace the damaged bridge, a substitute floating bridge from individual modules, suitable for the organized crossing is assembled [4]. Floating modules are used in conditions of natural disasters, e.g. during a flood. Undoubted advantages of this type of construction include the simplicity of the structure and the ability to organize various construction shapes depending on the water crossing characteristics. The advantage of floating bridges is also a possibility to use them regardless of the water obstacle depth and in various terrain conditions [1, 4].

The existing structural solutions of floating bridges are not able to meet the current requirements. Hence, a need to modernize crossing equipment emerges. Typical and outdate pontoon bridge constructions (e.g. PP-64 system) consist of a series of hermetic metal floating modules of large dimensions and complicated structure. Therefore, there are problems related to transport of well qualified operating teams. The restrictions to the usage of this type of heavy structure also result from operating conditions in a demanding environment. The use of steel pontoons with large dimensions requires adequate storage conditions and a sufficiently large storage space. These requirements are fulfilled by the structure of an innovative ribbon

Experimental Mechanics of Solids
Materials Research Proceedings **12** (2019) 96-103

Materials Research Forum LLC
https://doi.org/10.21741/9781644900215-14

assembled from a floating cassette pontoon bridge, proposed by the team of the Department of Mechanics and Applied Computer Science of the Military University of Technology, [4, 5, 6, 7].

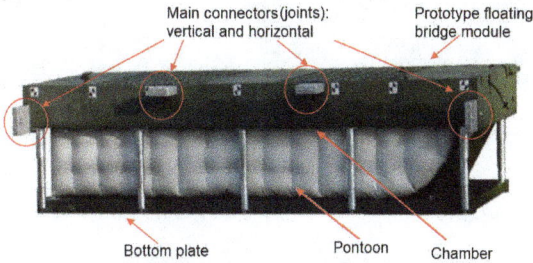

Fig. 1. *A module of a prototype floating bridge with the location of main joints set vertically and horizontally on a side of the construction*

Figure 1 shows a prototype floating cassette in which the bottom plate can be moved by pressure of a flexible pontoon operated inside a chamber on the bottom of the cassette. The analysed floating cassettes can be used, as all types of floating systems, in the form of footbridges, bridges, rafts or various types of jetties and quays. For special applications for the needs of an army, the cassettes should be assembled in the ribbon bridge. An unquestionable advantage of the cassette is regulation of the submersion which provides a possibility to leave the connected floating bridge below the water surface, which is the advantage of covering up the crossing a cross a river for, e.g., an enemy army. A re-use and return of the bridge to the water surface requires filling a flexible pontoon with the compressed air. A bridge built of pontoon cassettes can be assembled quickly and used as a substitute crossing organized in places where fixed bridges are not in service.

The main objective of this paper is to investigate strain maps of the floating bridge joint element with a usage of optical deformation measurement system. Similar investigations were carried out in [8], where the strength of a separated subsystem of a wagon for transport of trucks semitrailers is evaluated. The designed wagon allows easy and independent loading, transport and unloading without any special equipment or an additional platform infrastructure. The tests presented in paper [8] concern a wagon separated construction element – a side lock the most strained part of the wagon. Owing to the application of a non-contact optical system of strains measurement, the lock deformations as well as the areas of the minimum and maximum main deformations were defined.

Object of the research
The subject of the experimental research presented in this work is a separated main connector (joint) presented in Figure 2. There are four connectors on the both sides of each module of the floating bridge. Two are placed vertically and two – horizontally (Fig. 1). Side main connectors play an important role in the operation of the floating bridge [1, 4, 7]. They enable fast and easy connection of identical floating modules for assembling a complex ribbon of the bridge.

Experimental Mechanics of Solids
Materials Research Proceedings **12** (2019) 96-103

Materials Research Forum LLC
https://doi.org/10.21741/9781644900215-14

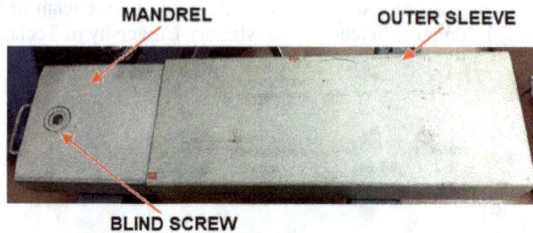

Fig. 2. The main joint of prototype floating bridge

Each connector consists of two components: a mandrel and an outer sleeve which has a closed rectangular profile. The mandrel is inserted into the sleeve of the adjacent bridge module. The blind screw is used to lock the connector elements of the adjacent module relative to each other.

The purpose of the research is to determine maps of deformation and strains on the one of the walls of the prototype connector sleeve. Using the FEM tools, the operating conditions of the joints connecting adjacent floating segments were simulated. It was established that the basic load, to which the structure is exposed, is modules. Therefore, the joint was subjected to a three-point bending. It was assumed in the research that the maximum stress occurring in the sleeve walls during loading cannot exceed the yield strength of the 18G2A steel, from which these elements are made of.

Loading conditions similar to those found inside connectors and in the roadway surface of real prototype cassettes were mapped during the lab tests with the use of the strength testing machine. The tests involved the usage of the Instron SATEC 1200 kN compressing machine, together with the installed three-point bending stand which is presented in Fig. 3a).

The distance between two bottom supports, with round half-shafts on the top, was equal to 945 mm. The mandrel rested on one half-shaft support and the sleeve rested on the other one. Because of different dimensions of the connector parts, half-shafts needed to have a different height, thus, the investigated connector was placed horizontally. Both half-shafts had radius of 29 mm. The upper punch had a radius of 140 mm. The mandrel was extended from the sleeve outer edge by 298 mm. Total length of the connector is equal to 1053 mm (Fig. 3b).

Three-point bending tests were carried out in the load-unload cycle until the maximum force of 100 kN was reached. The loading rate was equal to 10 mm/min. The behaviour of the joint in the configurations with and without the blind screw was compared in the research.

To determine the connector elements strains at selected points, the electroresistance method was used. A spot deformation obtained from the strain gauge marked with an arrow in Fig. 4b) was supplemented with the use of the optical 3D image correlation measurement system GOM Aramis (Fig. 4a). In this manner, maps with continuous distribution of deformations and strains were obtained. For this purpose, a stochastic black and white pattern was applied on one of the sleeve surfaces (Fig. 4b). The results of Aramis were verified by a strain gauge measurement method. The developed strain maps enabled precise, qualitative and quantitative, identification of the most stressed areas on the sleeve wall during bending of the joint and evaluation of the impact of the blind screw fixing mandrel in the sleeve (Fig. 4b) on the joint behaviour during loading.

Experimental Mechanics of Solids
Materials Research Proceedings **12** (2019) 96-103

Materials Research Forum LLC
https://doi.org/10.21741/9781644900215-14

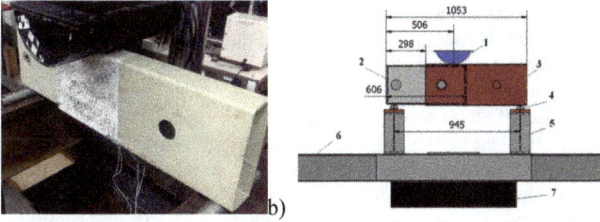

Fig. 3. The photo of the connector on the test stand (a) and the scheme of a three-point bending stand (b): 1- loading punch, 2 - mandrel, 3 - sleeve, 4 - half-shafts of the bottom support, 5 –adjusted support, 6 - I-beam of the stand, 7 –support of the testing machine

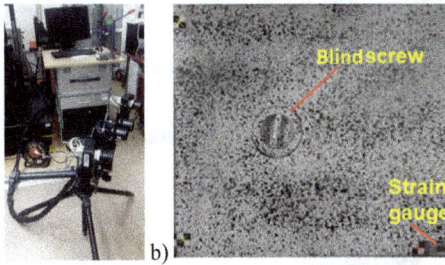

Fig. 4. Aramis optical deformation measurement system (a) and stochastic black and white pattern applied to the measuring part of the connector sleeve (b). The red arrow marks the blinds crew and the strain gauge attached near Point 2

Some areas of facets (the smallest area of the measurement in an optical deformation measurement system, equivalent to a strain gauge matrix) were marked in the Aramis software in order to obtain averaged strain curves in selected points of the sleeve. There were two points selected in the optical measurement results analysis for each evaluated configuration: one – near the blind screw hole marked as Point_1 and the other – near the strain gauge presented in Fig. 4 and marked as Point_2. The place corresponding to the strain gauge position could not be measured directly by the optical system. Therefore, comparison of both measurement methods – optical and electroresistance – is only approximate. The location of Points 1 and 2 are presented in Fig. 7 and Fig. 8 for both configurations, respectively.

Results of the research
During the load-unload cycles, hysteresis loops were recorded in the case of both the conducted test variants. The maximum loading force was equal to 100 kN. The connector without the screw achieved the assumed maximum force under deflection of 3.8 mm, while in the case of the fixed connector it was a higher value of 4.6 mm.

The results of testing the pontoon bridge joint in the configuration without the blind screw were obtained. Three photos from the test taken by the left camera of the 3D optical deformation measurement system are presented in Fig. 5. They show pictures of the joint without the screw: before the three-point bending test (a) at the maximum load of 100 kN (b) and at the end of the test (c). Generally, tests were conducted in the elastic region of the material, hence, the deformations are small. However, in some places a plastic deformation occurred.

The maps of main (major and minor) strains are shown in Fig. 6a) and b), respectively. Arrows of main directions are also shown in the pictures. The pressure of the loading stamp caused accumulation of stress near the upper surface of the connector sleeve. The area of strain concentration was marked with a red ellipse. The major strain of engineering strains in this area achieved a value of more than 0.23%. The black colour is used to mark the deformation concentration around the screw fastening hole. Within this area, two points of strain concentration occurred. They were slightly distant from the horizontal symmetric axis of the hole on both its sides – Fig. 6.

Fig. 5. Measurement area of the DIC system (marked in green) of the joint without a blind screw: a) before the test, b) at the maximum load, c) at the end of the test

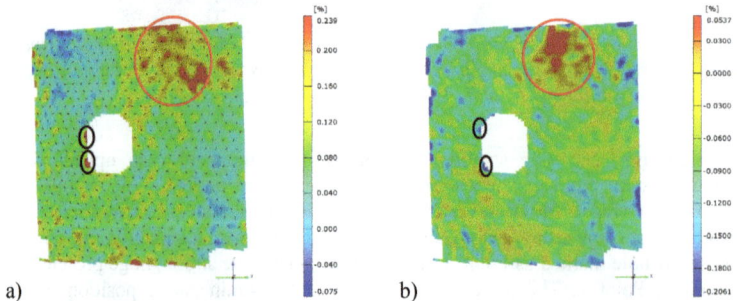

Fig. 6. Maps of the main strain distribution on the sleeve surface under maximum load for a joint with the blind screw: a) major and b) minor. Arrows of main directions are shown. Strain concentrations near the hole and near the stamp are marked with ellipses

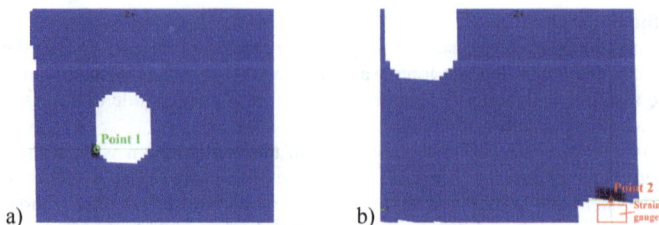

Fig. 7. The location of measurement points in the connector without a screw analysed in the 3D optical system: a) Point 1 near the screw hole; b) Point 2 near the strain gauge fixing

Experimental Mechanics of Solids   Materials Research Forum LLC
Materials Research Proceedings **12** (2019) 96-103   https://doi.org/10.21741/9781644900215-14

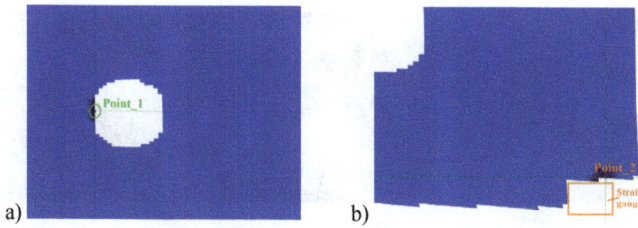

Fig. 8. The location of measurement points in the connector with the blind screw analysed in the 3D optical system: a) Point 1 near the screw hole; b) Point 2 near the strain gauge fixing

Fig. 7a) and b) presents the locations of Point 1 and 2 analysis on the optical measurement results for the joint configuration without a blind screw. Facets chosen for averaging the strains are marked in black. Point 1 is one of the points of strain concentration near the hole and Point 2 is slightly above the strain gauge measurement point marked with a red rectangular. Adequate points for the second joint configuration are presented in Fig. 8a) and b).

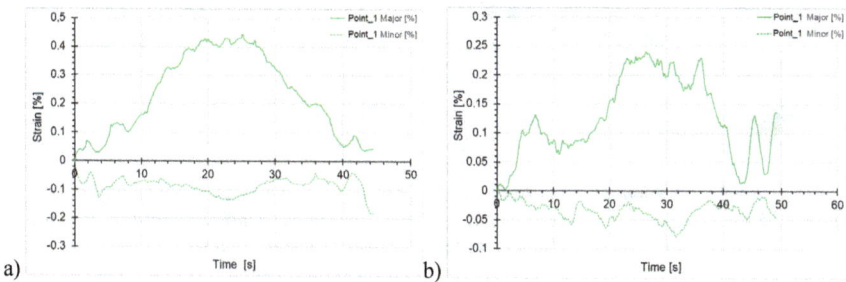

Fig. 9. Major and minor strains curves as a function of time for Point 1 (near the blind screw hole of the connector: a) without the screw; b) with the screw

In Fig. 9a) and b), a comparison of major and minor strains in Point 1 for both test scenarios is presented. In Point 1, near the hole without a screw, the maximum major strain averaged from the area marked in black in Fig. 7a) is equal more than 0.4%, which means that plastic deformations occurred in this area. In the second loading case, the maximum major strain for the area marked in black in Fig. 8a) was equal to 0.24%.

Similarly as in the case of the evaluations of the connector without a blind screw, the results of testing the pontoon bridge joint in the configuration with the blind screw are presented. Photos from the left camera of the DIC (Digital Image Correlation) system at the beginning, at the maximum load and at the end of the test are presented in Fig. 10. Next, maps of the main strains for the connector loaded with a maximum load of 100 kN are shown in Fig. 11. As in the first scenario, strain accumulation caused by the pressure of the loading punch (marked with red ellipses) occurred on the upper surface of the connector sleeve. The black colour ellipse is the place where the strains concentrated around the hole and the screw. There is only one point of the strain concentration in this scenario and it is situated approximately on the horizontal symmetry of the hole. The presence of the locking blind screw changed the character of the strain distribution in the area of the hole.

Experimental Mechanics of Solids
Materials Research Proceedings **12** (2019) 96-103

Materials Research Forum LLC
https://doi.org/10.21741/9781644900215-14

Fig. 10. Measurement area of the DIC system (marked in green) of the joint with the blind screw: a) before the test, b) at the maximum load, c) at the end of the test

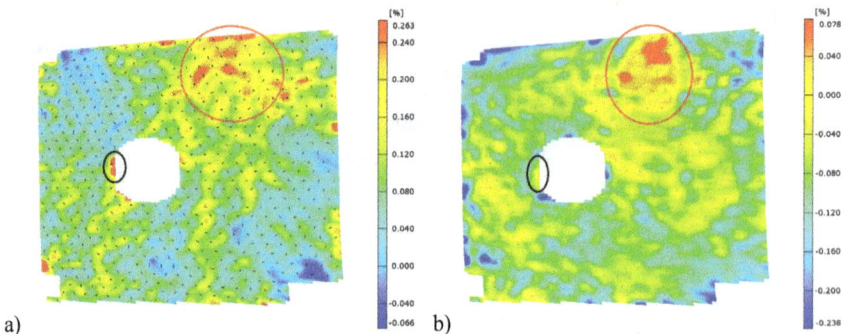

Fig. 11. Maps of the main strain distribution on the sleeve surface under maximum load for a joint with the blind screw: a) major and b) minor. Arrows of main directions are shown. Strain concentrations near the hole and near the stamp are marked with ellipses

Comparison between Points 2 chosen during DIC method and strain gauge measurements are presented for both scenarios in Fig. 12a)and b) respectively. The results compliance is sufficient but not exact. It should be noted that Points 2 were situated not exactly on the strain gauge fixing point but slightly above it. Moreover, strain gauge measurement is more precise compared with the optical measurement data.

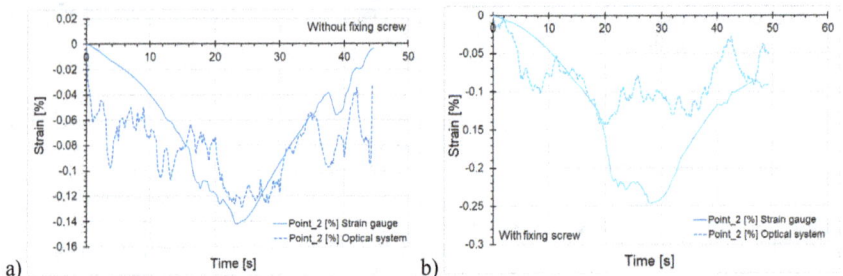

Fig. 12. Comparison of measurements carried out with the DIC system in Point 2 (dotted line) and the strain gauge near Point 2 (solid line) for the connector without (a) and with (b) the screw

Experimental Mechanics of Solids
Materials Research Proceedings **12** (2019) 96-103

Materials Research Forum LLC
https://doi.org/10.21741/9781644900215-14

Conclusions

Conclusions observed during three-point bending tests of the joint in configurations with and without a blind screw are as follows:

- The connector in both configurations generally worked in the elastic range of the material. However, there were areas where strain concentrations occurred in both cases. Measurements with the optical system allowed locating such places;
- The largest deformations were recorded near the place of the loading punch application on the upper surface of the sleeve and near the hole of the fastening bolt (Fig. 6 and 11);
- The nature of the deformations around the bolt hole is different for both scenarios. The presence of the blind screw interferes with the bending process and changes character of strain distribution especially near the hole. In the configuration without the screw, there are two independent foci. In the screw configuration, there is one place near the axis of symmetry;
- Maximum major strain in Point 1 is larger for configuration without the blind screw (Fig. 9a) than for the other configuration with the screw (Fig. 9b). The presence of the screw in the hole changes the nature of the connector load.

Acknowledgement

The works were carried out as part of the research project PBS 23-937, Military University of Technology, 2016-2018.

References

[1] W. Krasoń, Koncepcja, rozwiązania konstrukcyjne i badania systemu pływającego o regulowanej wyporności, rozdział w monografii. Inżynieria wojskowa - problemy i perspektywy, Wojskowy Instytut Techniki Inżynieryjnej im. profesora Józefa Kosackiego, Wrocław 2018, ISBN978-83-948983-0-4, pp. 123-146.

[2] W. Krason, J. Malachowski, Field test and numerical studies of the scissors-AVLB type bridge, Bulletin of The Polish Academy of Sciences, Technical Sciences, Vol. 62, No. 1, 2014, pp. 103-112. https://doi.org/10.2478/bpasts-2014-0012

[3] W. Krason, L. Filiks, Numerical tests of the main pin joint in scissor BLG bridge, CMM-2011 – Computer Methods in Mechanics, 9–12 May 2011, Warsaw.

[4] Derewońko, W. Krasoń, Mobile pontoon bridge and floating systems, Konferencja Antwerpia, maj 2018, Special Interest Group A2 (Ports and Maritime) of the World Conference on Transport Research Society (WCTRS), conference proceedings, 3-4 May 2018, Antwerpia.

[5] European Patent, No. 2251255, A sectional pontoon bridge, Military University of Technology, 2013.

[6] Patent Office of the RP, PAT.223689, Set of mechanical locks to connect the floating bridge cassettes and the cassette opening mechanism, Military University of Technology, 2016.

[7] W. Krason, P. Slawek, Design and pre-testing of a mobile modular floating platform with adjustable displacement, Mechanik nr 11, 2017, pp.1075-1080. https://doi.org/10.17814/mechanik.2017.11.185

[8] W. Barnat, W. Krasoń, P. Bogusz, M. Stankiewicz, Experimental and numerical tests of separated side lock of intermodal wagon, Journal of KONES Powertrain and Transport, Institute of Aviation, vol. 21, no 1, pp.15-22. https://doi.org/10.5604/12314005.1134049

Experimental Mechanics of Solids   Materials Research Forum LLC
Materials Research Proceedings 12 (2019) 104-109   https://doi.org/10.21741/9781644900215-15

Experimental Study into the Torsional Friction between AGV Wheel and Various Floors

Tomasz Bartkowiak[1,a*], Wojciech Paszkowiak[1,b], Marcin Pelic[1],
Adam Myszkowski[1]

[1]Poznan University of Technology, Institute of Mechanical Technology, Pl. M. Sklodowskiej-
Curie 5, 60-965 Poznan, Poland

[a]tomasz.bartkowiak@put.poznan.pl, [b]wojciech.z.paszkowiak@doctorate.put.poznan.pl

Keywords: Torsional Friction, Coefficient Of Friction, Hysteresis, Wheel Turn

Abstract. This paper presents an outcome of the experimental study into torsional friction between small sized wheels, intended for AGV, while being in contact in three different floors. The dedicated test stand was developed, which allowed testing a wheel while turning at various velocities. Each wheel was loaded via a pneumatic actuator. Turning was performed by parallel mechanism by using a turntable powered by a stepper motor. Nine different wheels were tested at three different turning angular velocities. Linear characteristics were obtained between torque required to turn and normal load, which allowed estimation of the coefficients of torsional friction for each wheel. A clear material hysteresis while turning and returning was observed for most cases. It was found that hardness and wheel geometry play important roles in the torsional behavior under load.

Introduction

The aim of this work is to study experimentally torsional behavior between AGV (Automated Guided Vehicle) wheel and typical floors. Different driving methods exist for four-wheeled vehicles: Ackermann steering, differential driving method, steering by using omnidirectional wheels or by skidding. Each of these approaches have certain advantages and disadvantages which determine their applicability in certain work conditions. They also implicate the dynamics of the vehicle during straight drive or turning, including frictional behavior. The subject of our study are commercially available wheels which are installed on a vehicle with the Ackermann steering system. In this case, apart from rolling friction, torsional friction occurs which can change with steering angular velocity.

Friction is one of the fundamental phenomena existing in nature. It significantly affects the operation of a mechanical system. The friction can be undesirable or desirable as it depends on the character and purpose of the design or process. An important dependence of contacting bodies is a relation of friction force as a function of the relative speed during contact [1]. In the basic engineering calculation it can be assumed that this relationship is constant. Stribeck proved that as the relative velocity of the bodies increases, the friction force decreases. This applies to the range of low slip velocities and is especially important when initiating stick-slip movements. Dry friction has discontinuities in its description, which leads to the solution being the result of approximation [1, 2, 3]. Interaction friction models, such as for example mass, spring and dashpot lead to the appearance of hysteresis friction model. Excessive friction contributes to wasting energy converted, finally, into financial losses. On the other hand, insufficient friction is the cause of many accidents. In production processes, the friction is vital for grinding and polishing [4].

In the literature, the common problem is the estimation of the car tire friction coefficient. Most often these studies concern lateral movement, longitudinal movement or their combination.

Experimental Mechanics of Solids
Materials Research Proceedings **12** (2019) 104-109

Materials Research Forum LLC
https://doi.org/10.21741/9781644900215-15

Each of these directions is associated with a unique value of coefficient of friction [5]. The lateral friction coefficient of a tire is the subject of multiple scientific papers. The main reason for this is unfavorable lateral tire deflection. The basic ground in the tests is usually asphalt [6]. Aligning moment occurring in these tests is not a torsional moment causing torsion friction. Despite that, its dependence in the displacement function occurs in the form of hysteresis [7].

Torsional contact is one of the main problems regarding the contact of bodies. The model assuming perfect smoothness of the contacting bodies leads to simplifications. In this model, two zones exist: slip area and adhesion area [2, 8]. Torsional moment is dependent on rolling speed. For maneuvers at a low speed, this torque takes maximal values. As the rolling speed increases, the steering torque decreases. It is also observed that the hysteresis loop is wider at lower speeds. The influence of the speed on the torque value was simulated using the finite element model. The simulation results were validated for a speed of 10 km/h by the vehicle steering torque tests [9]. In passenger vehicles, simulation procedures are used to show steering behavior during a turn. There are definitely more factors involved in this analysis than in the case of the model in which the torsional moment causes direct motion. The moment in the steering wheel occurs in the form of hysteresis [10]. The analyzes were devoted to road tires. A comparison of different types of wheels has not been carried out [5, 6, 7, 8, 9].

The torsional friction is a key factor in torsional fretting. The moment occurring in that process in the function of the steering angle is presented in the form of the aforementioned hysteresis. Hysteresis loop occurs for both metals and plastics [11, 12]. These studies have been extended showing the influence of the atmosphere on the hysteresis shape [13]. Experiments were conducted in order to check the influence of the number of rotation cycles on the change of the shape of the torque curve as a function of the angle of rotation. In this study, the influence of the pressure force and angular displacement amplitude on the form of hysteresis was also checked. It has been shown that the shape of the curve depends on the angle of torsion. For contacting materials PTFE, steel and normal load equal 123N hysteresis was rectangular for largest torsional angle(15° and 30°). For smaller steering angle(0,5° and 1°) shapes of curves were quasiparallelogrammatical. As the steering angle increases, the maximum torsional friction torque decreases until the angle reaches 5°. It has been shown that the increase of the normal force does not significantly affect the character of the moment curves, only the increase in amplitude. This was done for a narrow range of normal loads and only one pair of contact materials. For other data, parameter changes may have a different effect on the hysteresis shape [8]. Hysteresis also occurs in the frictional torsion dampers as dependence of the friction torque as a function of the torsion angle [14, 15]. On the basis of studies on torsional friction between the materials of an artificial knee joints, it was shown that material wear decreases with the increase in the steering angle amplitude, but increases with the pressure force. The torsional friction was simulated in ANSYS for rubber wheel using the finite element model. The amplitude of the rotation angle was ±2° [16].

Materials and methods
In the study, we used nine commercially available wheels of the same outside diameter (160 mm) and mounting hole (Fig. 1).. They are dedicated for internal transportation systems. The maximum load on the wheels from weight is between 250 and 400 kg. Material specifications of the investigated items are shown in Table 1. Specimen differ mostly in tread and tire material and resulting Shore hardness that is declared by the manufacturer. The core of wheel 6 was made of polyamide, the others were made of casted aluminum. The tire of wheel 7 is barrel-shaped which, according to the producer, is supposed to improve the torsional performance.

Experimental Mechanics of Solids   Materials Research Forum LLC
Materials Research Proceedings **12** (2019) 104-109   https://doi.org/10.21741/9781644900215-15

Fig. 1. Overview of the investigated wheels

In order to study the torsional phenomena, a dedicated test stand was developed (Fig. 2), which allowed testing a wheel while turning at various velocities. Each wheel was loaded normally via pneumatic actuator. Four load values were used: 1, 2, 3 and 4 kN. In order to maintain constant pressure and, thus, force, pressure was measured and stabilized through electrically powered pressure valve that was closed-loop controlled. Turning was performed by parallel mechanism and using turntable powered by stepper motor. All wheels were tested at three different turning angular velocities 10, 20 and 25 degrees per second. The material, in contact were concrete and two resin type floors (white – harder and blue – softer).

The torque required to turn the wheel under load was measured through a force meter fitted in the turning link. The torque was calculated taking into account the angular position. The data was collected by HBM DAQ unit connected to PC with Catman software installed. We have collected both the load force and torque. The turn angle was collected from a stepper motor controller.

Table. 1. Parameters of analyzed wheels.

Wheel number	Core material	Tread and tire material	Tread and tire hardness [° Shore A]
1	Aluminium	Extrathane® Polyurethane	92
2	Aluminium	TR® Polyurethane	95
3	Aluminium	Besthane® Polyurethane	92
4	Aluminium	EasyRoll® Rubber	65
5	Aluminium	Besthane® Soft Polyurethane	75
6	Polyamide	Besthane® Soft Polyurethane	75
7	Aluminium	TR-ROLL® Polyurethane	75
8	Aluminium	TR-ROLL® Polyurethane	75
9	Aluminium	Softhane® Polyurethane	75

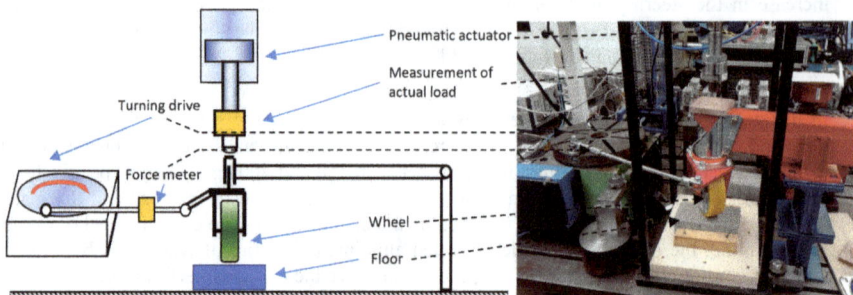

Fig. 2. The schematic and the actual view of the test stand

Experimental Mechanics of Solids   Materials Research Forum LLC
Materials Research Proceedings **12** (2019) 104-109   https://doi.org/10.21741/9781644900215-15

Results and discussion

The torque (torsional moment) versus turn angle relation took the form of hysteresis, which varied in shape depending on the wheel, floor and testing conditions. For each measurement we analyzed the four characteristic parameters: positive and negative remanence as well as positive and negative coercivity (Fig. 3).

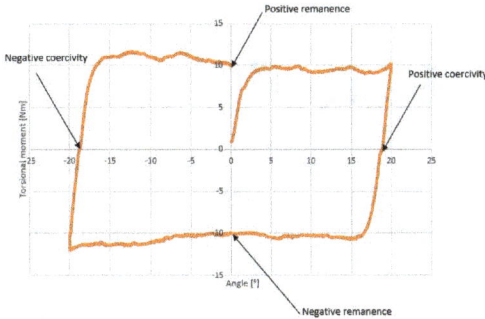

Fig. 3. Registered relation between torque and angle and hysteresis parameters.

We noticed that the shape of the hysteresis is mostly asymmetric and there are differences between positive and negative values of corresponding parameters. Those differences generally increase with the load, which is clearly evident for rubber tires (specimen 4) where the variations took the greatest values. The lowest differences in coercivity were observed for harder material (specimen 1-3). For the remanence no clear similar tendency could be noticed. We also found that the turning angular velocity (within 10-25 deg/s range) does not significantly influence the asymmetry of hysteresis. The results of mean differences between positive and negative coercivity as a function of load for all nine wheels while in contact with hard resin floor are shown   in Figure 4. We observed no buckling of the connecting rod (calculated force in the rod was below critical force), which could be a potential reason of hysteresis asymmetry.

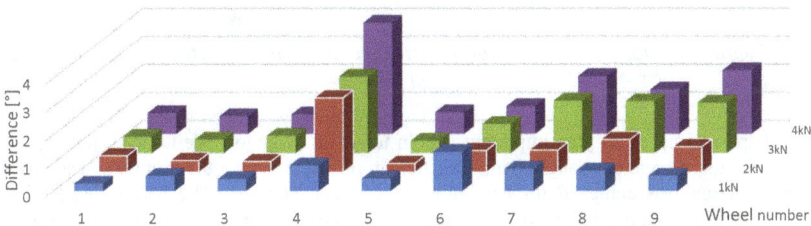

Fig. 4. Differences between positive and negative coercivity for all tested wheels

Basing on the obtained data, we determined characteristics between the torque required to turn and normal load, which allowed estimating coefficients of torsional friction for each wheel. We received a satisfactory linearity between load and torque ($R^2 > 0.9$ for linear regression). The characteristics for all measured wheels and floors are shown in Figure 5. We noticed that the highest values of torsional friction coefficient were measured for concrete paving floor. For that floor the wheel type does not influence the friction coefficient significantly when compared with other resin floors. Generally, greater values were observed for return than turn. In all cases, the rubber wheel showed the highest values of torsional friction coefficient.

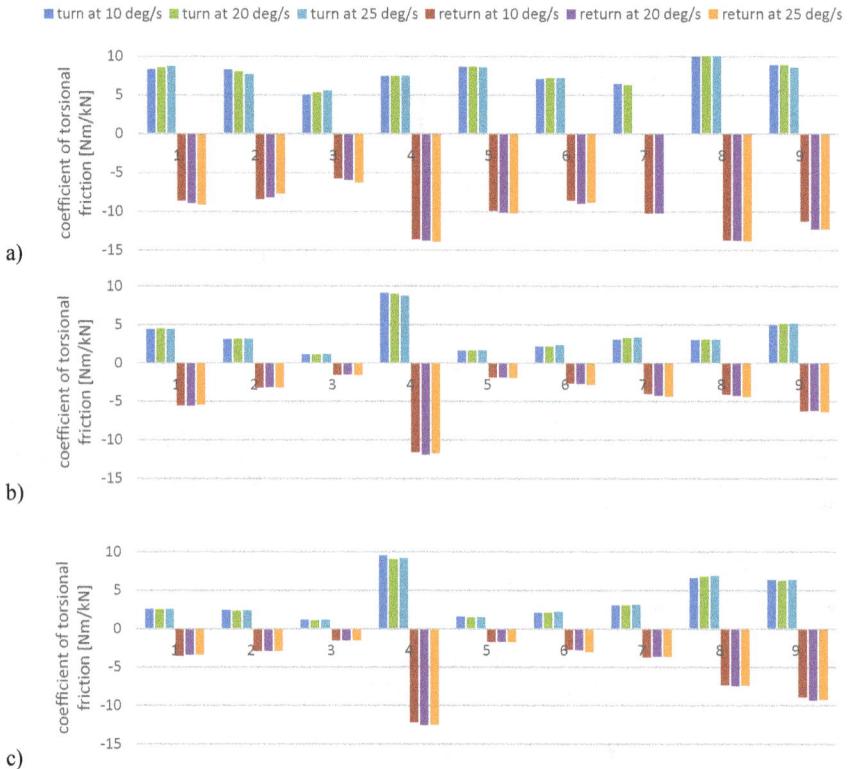

Fig. 5. Coefficient of torsional friction for all tested scenarios: a) concrete paving, b) hard resin floor, c) soft resin floor

Summary

Wheel shape and hardness play important role in torsional behavior. We noticed that hardness alone does not correlate strongly with coefficient of friction ($R^2 < 0.5$ utmost). The lowest torsional friction was achieved for hard material of tread and tire. The barrel shape wheels performed better when turning as compared to the wheels of the same material but flat tread. The rubber wheel required the most torque to turn in all tested conditions. The friction was also dependent strongly on the contacting material of the floor.

The estimation of the exact coefficient of friction is an important issue in the design of turning mechanism of an AGV. It determines the power of the drive, gearbox, battery size and electric current consumption. The torsional coefficient of friction can be difficult to estimate through simulation; therefore, a method for measuring the values in specific work condition might be a credible approach.

Experimental Mechanics of Solids
Materials Research Proceedings **12** (2019) 104-109

Materials Research Forum LLC
https://doi.org/10.21741/9781644900215-15

Acknowledgement
This research was supported by ATRES Sp. z o.o., as a part of the project entitled: "Opracowanie nowego typu wózka logistycznego oraz metody bezkolizyjnej i bezblokadowej realizacji procesów intralogistyki" (POIR.01.01.01-00-0485/17-00) funded by the National Centre for Research and Development, Poland.

References

[1] R. Cross, Increase in friction force with sliding speed, Am. J. Phys. 73 (2005) 812-816.

[2] F. Al-Bender, Fundamentals of friction modeling, Proceedings ASPE Spring Topical Meeting on Control of Precision Systems 48 (2010) 117-122.

[3] M. Wiercigroch, A note on the switch function for the stick-slip phenomenon, Journal of Sound and Vibration 175 (1994) 700-704. https://doi.org/10.1006/jsvi.1994.1559

[4] B. Drincić, Mechanical Models of Friction That Exhibit Hysteresis, Stick-Slip, and the Stribeck Effect, PhD Thesis, University of Michigan, 2012.

[5] H.B. Pacejka, Tyre and Vehicle Dynamics, second ed., Butterworth-Heinemann, 2006.

[6] S. Hong, J.K. Hedrick, Tire-Road Friction Coefficient Estimation with Vehicle Steering, IEEE Intelligent Vehicles Symposium IV (2013) 1227-1232. https://doi.org/10.1109/ivs.2013.6629634

[7] Y. Li, J. Zhang, X. Guan, Estimation of Vehicle Parameters and Road Friction Using Steering Torque and Wheel Speeds, WSEAS Transactions on Systems 11 (2012) 1-11.

[8] S. Wang, C. Niu, Torsional Tribological Behavior and Torsional Friction Model of Polytetrafluoroethylene against 1045 Steel, PLOS ONE 10 (2016) e0147598. https://doi.org/10.1371/journal.pone.0147598

[9] Y. Wei, C. Oertel, Y. Liu, X. Li, A theoretical model of speed-dependent steering torque for rolling tyres, Vehicle Syst. Dyn. 54 (2016) 463-473. https://doi.org/10.1080/00423114.2015.1111391

[10] J. Loof, I. Besselink, H. Nijmeijer, Component based modeling and validation of a steering system for a commercial vehicle, The Dynamics of Vehicles o Roads and Tracks IAVSD 2015 (2016) 15-24. https://doi.org/10.1201/b21185-4

[11] Z. Cai, M. Zhu, Z. Zhou, An experimental study torsional fretting behaviors of LZ50 steel, Tribology International 43 (2010) 361-369. https://doi.org/10.1016/j.triboint.2009.06.016

[12] J. Yu, Z. Cai, M. Zhu, S. Qu, Z. Zhou, Study on torsional fretting behavior of UHMWPE, Applied Surface Science 255 (2008) 616-618. https://doi.org/10.1016/j.apsusc.2008.06.179

[13] Z. Cai, M. Zhu, J. Zheng, X. Jin, Z. Zhou, Torsional fretting behaviors of LZ50 steel in air and nitrogen, Tribology International 42 (2019) 1676-1683. https://doi.org/10.1016/j.triboint.2009.04.031

[14] Z. Skup, Structural friction and viscous damping in a frictional torsion dumper, Journal of Theoretical and Applied Mechanics 40 (2002) 497-511.

[15] Z. Skup, Damping of vibrations in a power transmission system containing a friction clutch, Journal of Theoretical and Applied Mechanics 43 (2005) 875-892.

[16] H. Xu, K. Chen, D. Zhang, X. Yang, Torsional friction of the contact interface between the materials of an artificial knee joint replacement, Journal of Biomaterials Science 29 (2018) 562-581. https://doi.org/10.1080/09205063.2018.1426921

Experimental Mechanics of Solids
Materials Research Proceedings 12 (2019) 110-116

Materials Research Forum LLC
https://doi.org/10.21741/9781644900215-16

Issues of Load Identification Using an Integrated Forces and Torques Sensor

Piotr Dudziński[a], Robert Czabanowski[b,*], Adam Konieczny[c], Andrzej Kosiara[d], Aleksander Skurjat[e] and Damian Stefanow[f]

Department of Off-Road Machine and Vehicle Engineering, Wrocław University of Science and Technology, Łukasiewicza 7/9, 50-371 Wrocław, Poland

[a]Piotr.Dudzinski@pwr.edu.pl, [b]Robert.Czabanowski@pwr.edu.pl,
[c]Adam.Konieczny@pwr.edu.pl, [d]Andrzej.Kosiara@pwr.edu.pl, [e]Aleksander.Skurjat@pwr.edu.pl,
[f]Damian.Sefanow@pwr.edu.pl

Keywords: Experimental Identification, Six-Axis Load Sensor, Calibration, Excavator, Finite Element Method

Abstract. The article presents a number of problems resulting from the need to identify complex loads acting on a working tool of a one-bucket excavator. The paper is focused on the presentation of conceptual design, calibration and protection of six-axis load sensor. A finite element model has been used to assess the applicability of measuring elements in a standard quick-coupler.

Introduction

Modern work machines (e.g. excavators, loaders) are used for various tasks, which require the use of various work tools. In order to optimally utilize the working time of the machine, quick-couplers are often used for excavators or loaders work tools, enabling quick and reliable tool changes, even without leaving the cabin by the operator. This tool change system is possible using hydromechanical quick-couplers controlled from the cabin. The use of various work tools is associated with wider requirements regarding the working system of the machine and its control system. It is worth mentioning here the need, for example, to limit the available force on the edge of the tool depending on the type and size of the tool. Implemented control systems make it possible to: monitor the loads acting on the tool and weigh the excavated material. It should be emphasized that permanent load control allows for: protection of the tool and work tools against overload or destruction, measurement of the excavated mass in bucket or scoop and performance evaluation, control of machine stability, evaluation of effort and forecasting tool durability. Therefore, as part of the European grant PROSYMA (in cooperation with the following partners: Cologne University of Applied Sciences, Lehnhoff Hartstahl GmbH & Co. KG, Gunderson & Løken AS and Sensors and Synergy SA), a system for identification of work tool loads was developed [1].

The concept of a sensor for measuring of force and torque

As part of the implemented project, the own concept of force and torque sensor was developed and implemented because the products available on the market could not, mainly due to the large dimensions, be installed in a standard quick coupler. The Lehnhoff Hartstahl quick-coupler was used after modification without affecting the functionality of the quick-coupler and machine performance (Fig. 1). The modification of the standard VL210 quick-coupler consisted in obtaining sufficient space for the assembly of the measuring segments system [1].

The six-axis load sensor was designed as a system of 4 measuring beams (Fig. 1-4) made of high-strength steel. Such material was used to obtain the highest possible sensitivity while

Experimental Mechanics of Solids   Materials Research Forum LLC
Materials Research Proceedings **12** (2019) 110-116   https://doi.org/10.21741/9781644900215-16

maintaining the ability to carry large loads (the VL210 quick coupler is dedicated to machines with a working weight of 20 t, which gives components of forces above 200 kN). In order to determine the potential of the measurement elements by determination of their strain and to verify the correctness of introduced structural changes to the original quick-coupler design, numerical simulations using finite element method were conducted (Fig. 2).

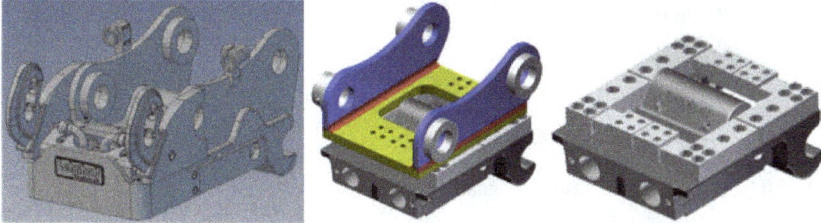

Fig. 1. Modification of the standard VL210 quick-coupler

The sensor of force and torque was designed as a system of 12 galvanic separated full strain gauge Wheastone bridges (Fig. 3, 4).

Fig. 2. Exemplary of results of numerical calculations of a measuring beams

Due to the fact that each of the measurement segments contained strain gauge systems dedicated to identification of individual load components, a sensor characteristic was developed to take into account the interaction between individual load components. Bearing in mind the previous experience in the construction and operation of multidimensional measuring sensors [1, 2], a matrix form of the characteristic was adopted (equations 1).

The matrix form of multi-axis sensor characteristic was adopted assuming that relations between force/moment and strain are linear, and that the relations between strain and voltage output are linear too [2, 3, 4].

$$
\begin{bmatrix} F_x \\ F_y \\ F_z \\ M_x \\ M_y \\ M_z \end{bmatrix} = \begin{bmatrix} a_{11} \ a_{12} \ a_{13} \ a_{14} \ a_{15} \ a_{16} \\ a_{21} \ a_{22} \ a_{23} \ a_{24} \ a_{25} \ a_{26} \\ a_{31} \ a_{32} \ a_{33} \ a_{34} \ a_{35} \ a_{36} \\ a_{41} \ a_{42} \ a_{43} \ a_{44} \ a_{45} \ a_{46} \\ a_{51} \ a_{52} \ a_{53} \ a_{54} \ a_{55} \ a_{56} \\ a_{61} \ a_{62} \ a_{63} \ a_{64} \ a_{65} \ a_{66} \end{bmatrix} \begin{bmatrix} U_{F_x} \\ U_{F_y} \\ U_{F_z} \\ U_{M_x} \\ U_{M_y} \\ U_{M_z} \end{bmatrix} \quad \text{or} \quad [\bar{O}] = [\bar{C}][\bar{U}]
$$
(1)

where:

[F_x..M_z] =[$\bar{O}$] – load vector,
[a_{11}..a_{66}] = [$\bar{C}$] – matrix of sensitivity factors,
[U_{Fx}..U_{Mz}] = [$\bar{U}$] – measuring signals vector.

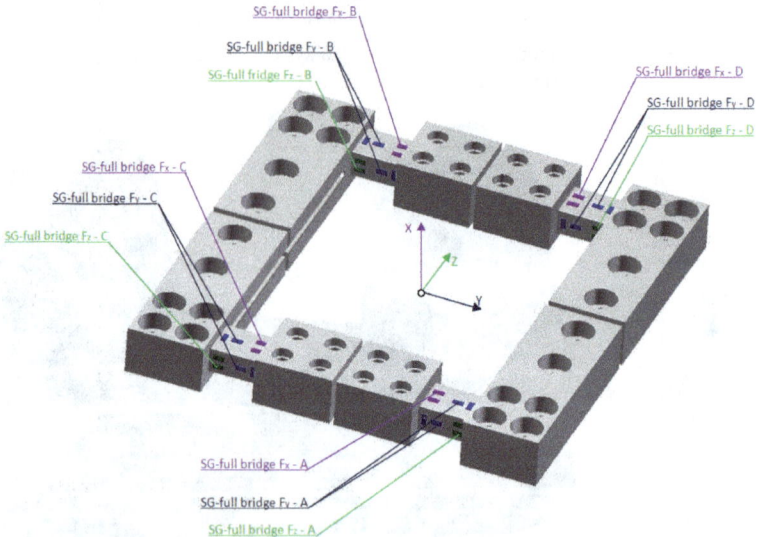

Fig. 3. Measurement beams system for identification of force-torque components

Calibration of six-axis load sensor

The complex form of the characteristic (equations 1) required proper conduct of experimental scaling of the measurement system. In order to obtain not peculiar matrix of coefficients, experimental testing was carried out by loading the system with six independent load cases. The Hottinger Baldwin Messtechnik QuantumX MX1615 amplifier, operating under control of CATMAN software was used for scaling. The testing was performed in two stages: in laboratory and in-situ tests.

To carry out laboratory pre-calibration laboratory strength testing machines were used. Due to the limited space, the tested system of measuring segments was mounted in specially designed housings enabling application of required load conditions using a testing machine. Figure 5 shows the arrangement of measuring segments at selected load states carried out during this stage of testing.

Experimental Mechanics of Solids
Materials Research Proceedings **12** (2019) 110-116

Materials Research Forum LLC
https://doi.org/10.21741/9781644900215-16

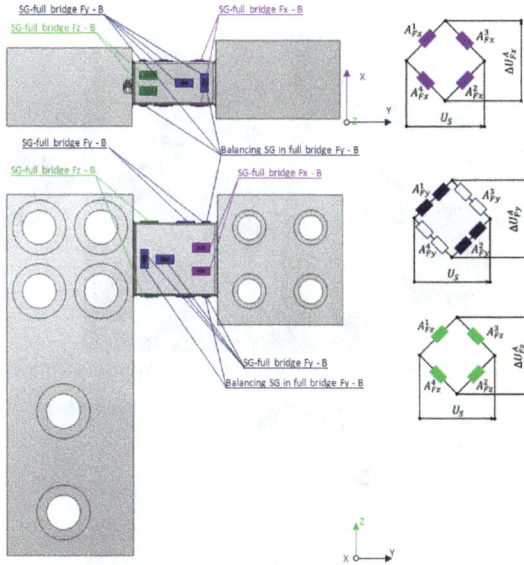

Fig. 4. Connections of strain gauge bridges at one of the measuring beams

The obtained results positively verified the concept of measuring force and torque, but due to the specificity of measuring segments assembly (location of measuring segments in a modified quick-coupler, required clamping forces), the obtained characteristic was treated as a preliminary, requiring verification on the work machine.

The final scaling of the force and torque sensor was carried out during in-situ tests. The system of measuring segments was mounted in the modified VL210 quick-coupler. The tests of the complete assembly were carried out on a Komatsu PC210 excavator with a mounted bucket (Fig. 6). Measuring scale was used to verify the real values of forces acting on the work tool.

Like in the laboratory, the tests were carried out in a manner enabling obtaining the required number of independent load states (two examples are shown in Figure 7). The obtained results of the measurements allowed, after statistical processing of a large set of results, to obtain the value of 36 coefficients of the sensitivity matrix (vide equation 1):

$$[\bar{C}] = \begin{bmatrix} 32,845 & -140,187 & -7,020 & 76,015 & 10,092 & 10,429 \\ 1,626 & 317,297 & 0,538 & 156,789 & -8,246 & -13,245 \\ -5,388 & 25,936 & -6,712 & -63,851 & 18,714 & -14,175 \\ 2,182 & -62,398 & 3,246 & -18,024 & -6,218 & 9,463 \\ 8,812 & 12,338 & -0,871 & 134,629 & 41,229 & -21,936 \\ -3,489 & 235,199 & 24,029 & -615,501 & 2,322 & 112,450 \end{bmatrix} \tag{2}$$

The values of the impact coefficients (a_{ij}, for $i \neq j$) are different for the order of magnitude, this is mainly due to the different stiffness of the measuring segments in the individual directions.

113

Experimental Mechanics of Solids
Materials Research Proceedings **12** (2019) 110-116

Materials Research Forum LLC
https://doi.org/10.21741/9781644900215-16

Fig. 5. Laboratory pre-scaling of force and torque sensor loaded with F_z and F_x

Fig. 6. Komatsu PC210 excavator equipped with modified VL210 quick-coupler with force and torque sensor and bucket during in-situ tests

Protection of strain gauge sensor

All measuring devices applied in a working machine operating in a wide range of environmental parameters must be properly secured. In the tested prototype it was not possible to use a closed housing. Due to the working conditions during the planned tests strain gauges and cable connections were secured with covering agent PU140 and covering putty AK22 and aluminum foil (products of Hottinger Baldwin Messtechnik).

During all tests (Fig. 8), the applied protective coatings ensured the required level of protection against external factors (temperature, pollution, humidity), but in the case of applications in more difficult conditions (impacts of rock fragments, presence of large amounts of water), it would be necessary to rebuild the quick-coupler so that the sensor can be better protected.

Fig. 7. Final calibration of the sensor prototype allowing the measurement of six load components using bucket and measuring scale

Fig. 8. Measuring beams with protective coatings mounted in the modified quick-coupler during digging

Tests of multi-axis load sensor during normal operation

The measurement results (Fig. 9) obtained during digging (Fig. 8) show the correct operation of the force and torque sensor and the possibility of its application in the quick-coupler. In the diagram shown (Fig. 9) fluctuations in load indications (here F_x, F_y and F_z components) are predictable when unloading the bucket, whereas the mass calculated on the basis of the characteristics of the force and torque sensor significantly changes its values because the values shown in the diagram do not include corrections calculated by the system based on the measured accelerations.

Summary

A new force and moment sensor as a part of modified quick-coupler has been designed, built, calibrated and tested. In this paper some problems with multi-axis load sensor have been described. The use of the finite element method allows the verification of the initial concept of the form of the deformable part of the force and torque transducer.

Preparation for work of a strain gauge multi-axis sensor requires a particularly accurate determination of the characteristic, and in the case of predicted use in difficult condition, adequate protection.

The experimentally tested modified quick-coupler with the multi-axis load sensor can be implemented in machines using various work tools due to the use of quick-couplers. The use of

Experimental Mechanics of Solids
Materials Research Proceedings **12** (2019) 110-116

Materials Research Forum LLC
https://doi.org/10.21741/9781644900215-16

this solution allows to protect the machine and tools from overloads, evaluate machine performance and help protect the machine against loss of stability.

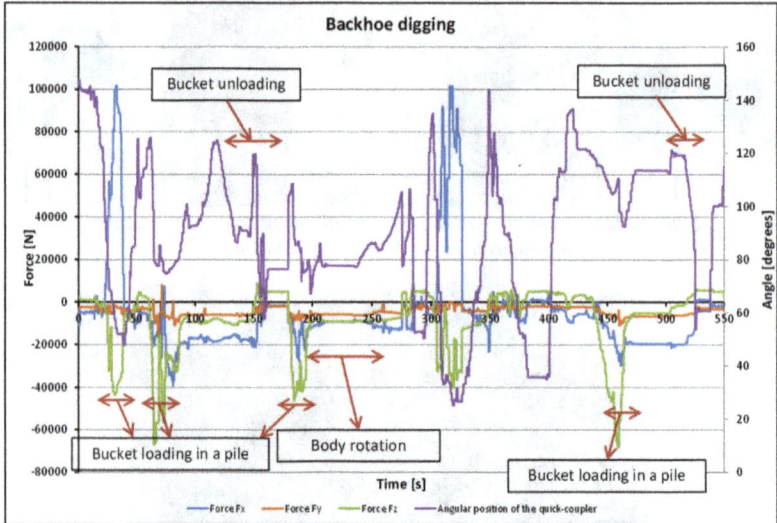

Fig. 9. Forces acting on the modified quick-coupler during digging process

Acknowledgement
The research was carried out as part of the PROSYMA project "Process-Optimized System Functionality of Mobile Work Machines", financed under the 7th Framework Program, (ID: 606227, FP7-SME-2013).

References

[1] M. Bürkel, B.E. Busterud, S. Sendrowicz, A. Ulrich, J. Lommatsch, M. Heuckeroth, M. Diefenbach, P. Dudziński, R. Czabanowski, A. Konieczny, A. Kosiara., A. Skurjat, D. Stefanow, Development of a modular information and assistance system for excavators, Fachtagung Baumaschinentechnik: Maschinen, Prozesse, Vernetzung: Tagungsband, Forschungsvereinigung Bau- und Baustoffmaschinen, Frankfurt, 2015, pp. 241-252.

[2] P. Dudziński, An unconventional transducer for measuring forces and moments in machine rotary pairs, 2nd Scientific Conference - Experimental Methods in Construction and Operation of Machines, Wrocław-Szklarska Poręba, 1995.

[3] G. Mastinu, M. Gobbi, G. Previati, A New Six-axis Load Cell. Part I: Design, Experimental Mechanics, Volume 51 (2011), p. 373-388. https://doi.org/10.1007/s11340-010-9355-1

[4] G. Mastinu, M. Gobbi, G. Previati, A New Six-axis Load Cell. Part II: Error Analysis, Construction and Experimental Assessment of Performances Experimental Mechanics, Volume 51 (2011), p. 389-399. https://doi.org/10.1007/s11340-010-9350-6

Experimental Mechanics of Solids
Materials Research Proceedings **12** (2019) 117-123

Materials Research Forum LLC
https://doi.org/10.21741/9781644900215-17

An Integrated Application to Support the Implementation of Measurements and Data Analysis

Robert Czabanowski

Department of Off-Road Machine and Vehicle Engineering, Wrocław University of Science and Technology, Łukasiewicza 7/9, 50-371 Wrocław, Poland

Robert.Czabanowski@pwr.edu.pl

Keywords: Measurement Application, LabView, Virtual Instrument, DAQ

Abstract. The article presents the design of an application created for the integration of various measuring devices in complex measurement systems. Uniform user interface with the necessary functionalities essential for the implementation of measurements simplifies carrying out large measurement tasks. The modular and open structure of the program allows for adapting to current needs. The applied LabView environment enables easy implementation of additional functions related to on-line processing of measured quantities.

Introduction

In the Laboratory of the Department of Off-Road Machine and Vehicle Engineering of Wrocław University of Science and Technology, as in many others, various control and measurement equipment is used. This diversity applies to producers, classes, supported (and most commonly used) communication interfaces as well as the functions that are most often used. The most frequently applied devices are: measuring amplifiers (often used for cooperation with strain gauge sensors), specialized or multifunctional measurement cards (DAQ), PLC controllers, digital multimeters and oscilloscopes, signal analyzer, and others (Fig. 1).

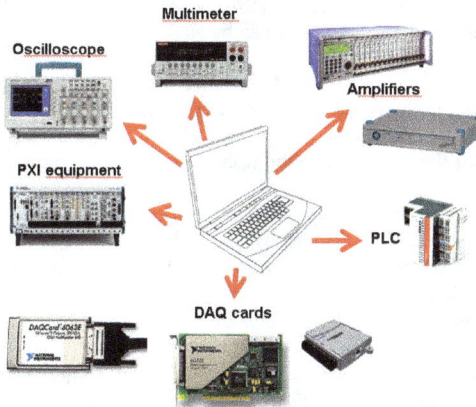

Fig. 1. Laboratory measuring devices with various interfaces (PCI/PXI, USB, FireWire, WiFi, RS232C, GPIB, Centronics, TCP/IP et al.)

Today these devices are mostly constructed so that they can be a part of a computerized measurement system even if it is possible to use the equipment without working with a

Experimental Mechanics of Solids   Materials Research Forum LLC
Materials Research Proceedings **12** (2019) 117-123   https://doi.org/10.21741/9781644900215-17

computer. To fulfill this task, hardware manufacturers offer dedicated software or at least drivers to enable the creation of custom applications (using e.g.: C, C ++, assembler, Python et al.). In the case of building a more complex measurement or control and measurement system, when the system designer is forced to use various devices, there are difficulties in system integration. Applications offered by hardware manufacturers can't deal well with the hardware support of competition. The functionalities offered by these programs are not always sufficient and require the development of special procedures for processing the data recorded in the off-line using other tools (Matlab, Statistica etc.) or also creating macros or extensions (e.g. by generating scripts or subroutines using VBA).

Requirements for a measurement software
For the needs of the implementation of a wide range of measurement tasks, a measurement application should meet the following requirements:
1. Possibility the use of various measuring devices using a uniform user interface.
2. Configuration of the measuring and control devices (measurement interface, synchronization method, type of trigger).
3. Configuration of the individual measurements channels (measurement range, sampling frequency, method of using the measurement signal).
4. Possibility of introducing the owned characteristics of measurement channels or of obtaining them using the teach-in method, in order to obtain the measurement results in the desired units.
5. The ability to perform on-line calculations on the values obtained from several measurement channels so that you can directly get measurements of quantities requiring data from several measurement channels.
6. On-line visualization of measurement results in a chosen way (several typical digital and graphic indicators and various diagrams).
7. The choice of how to use the measurement data, e.g.: archiving, off-line processing.

A lot of measurement programs offered on the market meet many of these requirements. The best of the programs used by the author, Catman by Hottinger Baldwin Messtechnik, is designed to support the products of this manufacturer and allows to service only some of the measurement cards (DAQ) of other manufacturers. In order to integrate devices from other manufacturers in measurement stands built in the Laboratory of the Department of Off-Road Machine and Vehicle Engineering a measurement program was created that allows for the operation of various measuring devices from the level of one application run on one computer operating in the MS Windows environment.

The measurement program was created using the National Instruments LabView package [1]. The choice of this environment was decided by:
- High versatility - many manufacturers of measuring equipment provide complete libraries or at least LabView drivers;
- A wide range of ready-made elements to create an intuitive and functional user interface;
- The ability to create applications using graphical programming;
- The author's previous experience in creating applications in LabVIEW.

Programming in the LabView environment is fundamentally different from coding in higher-level languages (C, C ++, Python, et al.). Applications (so-called "virtual instruments") are created in two mutually integrated parts: a front panel in which an interactive user interface is defined (equipped with digital and graphic indicators and controls for data entry or process control) and a system diagram that contains a program code represented by a system of connected function blocks that implement the desired algorithm. Of course, it is possible to use typical structures such as loops and conditional statements, as well as dividing applications into subprograms (they are defined as separate virtual instruments). This allows the construction of applications with specific functionalities: numerical calculations [1, 2], testing of products [1, 3], testing of a virtual PLC [1, 4].

Software developers who have procedures written in the above-mentioned higher-level languages can implement them in LabView. This programming environment allows the use of directly procedures developed using the Matlab package [1, 5].

The concept of integrated measurement application

The implemented application has a modular structure enabling the execution of a measurement task using the intuitive front panel (Fig. 2).

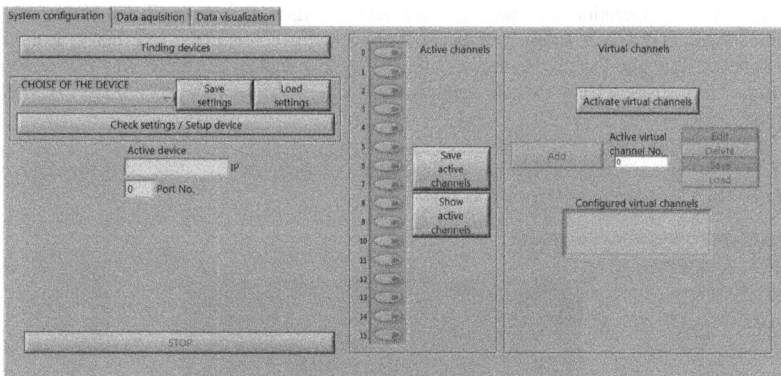

Fig. 2. Panel for devices configuration and set of individual measuring channels

The developed program enables:
1. Initialization communication with the selected measuring device ("Find device" button on main front panel - Fig. 2) - depending on the type of device, it is also possible to select the type of communication interface.
2. Checking the device settings and their correction - each initiated device can be individually configured both by manual correction of individual parameters as well as by loading the configuration saved in an external file using a different program option.
3. When operating multi-channel devices, the program allows setting the parameters of each channel:
   - setting the type of connected sensor,
   - determination of the measuring range and sampling frequency,
   - setup of the dimension of the measured quantity,
   - introduction of sensor characteristics (in this version of the program only by providing gain and offset),
   - specific low-pass filtration parameters (type and bandwidth) - this applies to measuring devices equipped with such filters,
   - setting the sensor power supply parameters from the internal power supply of the measuring device,
   - channel zeroing.
4. Record and read device configuration (to/from mass memory).

5. Activation and deactivation of selected measurement channels.
6. Support for virtual channels:
   Virtual channels were created to visualize (and register) on-line quantities that are the result of indirect measurements. Very often, the value of an interesting physical quantity is the result of more or less complex calculations on the values of signals from individual sensors. In addition to controlling and registering interesting physical quantities, such on-line calculations allow you to reduce the work on the development of test reports. The user can define any number of virtual channels using the right part of the front panel (Fig. 2). The limitation of the current version of the program is to support only a few basic mathematical functions (addition, subtraction, multiplication, division and exponentiation - also with the use of introduced constants). Although the LabView package allows the implementation of even complicated numerical calculations [2], in the presented version of the application, to simplify the operation of the equation editor (when defining virtual channels), the implementation of a larger number of mathematical functions has been abandoned. However, this allows for the introduction of more complex characteristics of the sensors used. To avoid rewriting virtual channels many times, the user can save them to a file and read.
7. Acquisition of measurement signals:
   A special operator panel (Fig. 3) was developed to configure this stage of the measurement task. The user has the following functions:

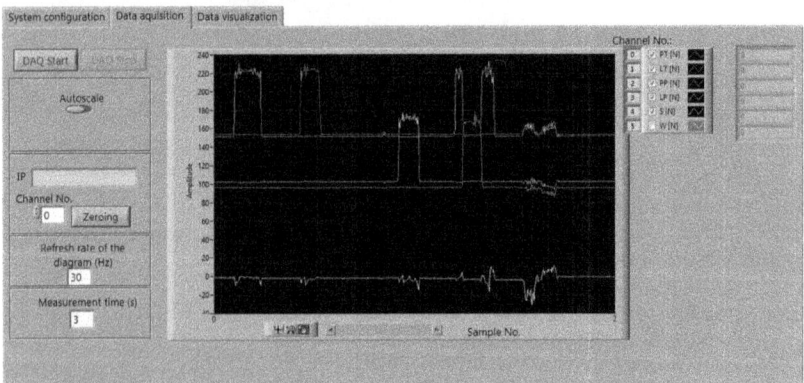

Fig. 3. Panel for acquisition of selected values measured during tests

- start and stop measurement,
- specify the duration of measurements,
- choose the diagram scaling method: automatic (adjusts the range of graphically displayed quantities), manual (allows to define the scope of the "Amplitude" axis on the graph),
- zeroing the selected channel,
- set the image refreshment rate of the diagram,
- the ability to activate and deactivate the display of each channel's value on the diagram,
- viewing numerical values of active channels,
- saving results to a file (the function is activated after stopping signal acquisition).

8. Visualization of measurement data:
   In the current version of the program, this is a functionality that requires manual loading of a file with a schematic or photo (supported formats: JPG, BMP) and deployments on digital indicators in the appropriate places on the image. By selecting the "Display settings" function, the user can define the format and range of the data displayed in the indicator and assign indicators to the measuring channels. Figure 4 presents the visualization prepared for the needs of the stand for testing the load process with a bucket of a one-bucket loader [6] in the Laboratory of the Department of Off-Road Machine and Vehicle Engineering.

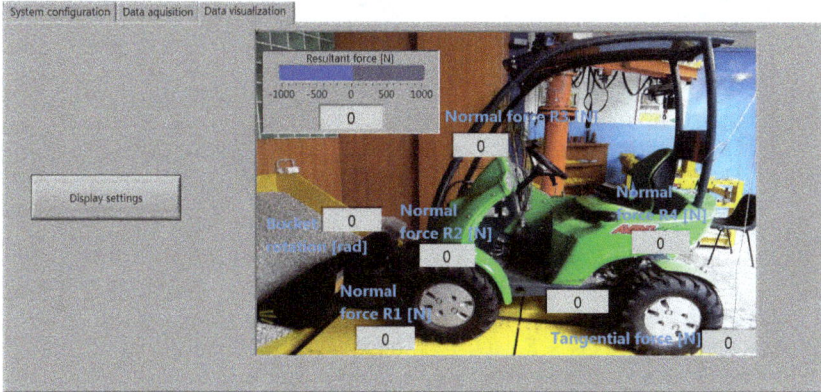

Fig. 4. Panel with an example graphic visualization for acquisition of selected measured quantities (channels)

The functions of the program described above can be implemented thanks to the activated subpanels that appear during the implementation of some program functions (Fig. 5).

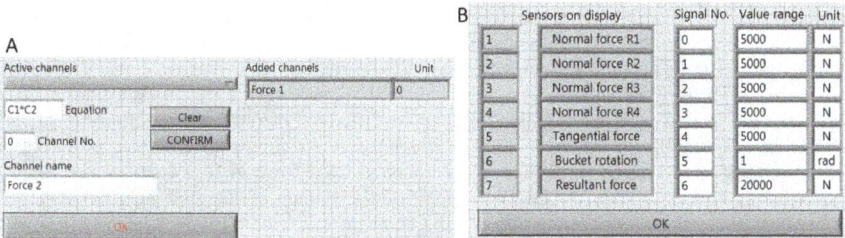

Fig. 5. Selected subpanels for detailed configuration of program functions (A - subpanel for define of virtual channels, B - subpanel for configuration of display settings

Software functionality tests on a laboratory stand

One of the important functional tests of the developed measurement program was carried out on a laboratory stand equipped with a series of sensors designed to identify the loads of the working tool (bucket), its position and wheel loads (Fig. 6). The loads are measured by strain gauges sensors (cooperating with the QuantumX amplifier), the position of the bucket is identified by potentiometric transducers (during normal operation, the potentiometers also cooperate with the QuantumX amplifier, here a simple DAQ card, NI USB-6009, was used for the test).

Experimental Mechanics of Solids
Materials Research Proceedings **12** (2019) 117-123

Materials Research Forum LLC
https://doi.org/10.21741/9781644900215-17

The use of virtual channels turned out to be particularly useful for cooperation with a measuring system equipped with a multi-dimensional sensor of forces and moments. The load sensor shown in Figure 6 makes it possible to identify 6 load components, 3 forces (Fx, Fy, Fz) and 3 moments (Mx, My, Mz). The developed matrix characteristic of the load transducer [6] (equations 1, 2) allows to identify the components of load, take (filter out) inter-channel crosstalk (between measurement channels), and also eg: calculate the resultant load. The use of virtual channels allows the introduction of all the 36 coefficients of the characteristics, calculating a desired quantities, the visualization and archiving. In Figure 4, the resultant force was calculated using virtual channels.

$$
\begin{bmatrix} F_x \\ F_y \\ F_z \\ M_x \\ M_y \\ M_z \end{bmatrix} = \begin{bmatrix} A_1 & A_2 & A_3 & A_4 & A_5 & A_6 \\ B_1 & B_2 & B_3 & B_4 & B_5 & B_6 \\ C_1 & C_2 & C_3 & C_4 & C_5 & C_6 \\ D_1 & D_2 & D_3 & D_4 & D_5 & D_6 \\ E_1 & E_2 & E_3 & E_4 & E_5 & E_6 \\ F_1 & F_2 & F_3 & F_4 & F_5 & F_6 \end{bmatrix} \begin{bmatrix} U_{Fx} \\ U_{Fy} \\ U_{Fz} \\ U_{Mx} \\ U_{My} \\ U_{Mz} \end{bmatrix} \tag{1}
$$

$$
\begin{cases} F_x = A_1 \cdot U_{Fx} + A_2 \cdot U_{Fy} + A_3 \cdot U_{Fz} + A_4 \cdot U_{Mx} + A_5 \cdot U_{My} + A_6 \cdot U_{Mz} \\ F_y = B_1 \cdot U_{Fx} + B_2 \cdot U_{Fy} + B_3 \cdot U_{Fz} + B_4 \cdot U_{Mx} + B_5 \cdot U_{My} + B_6 \cdot U_{Mz} \\ F_z = C_1 \cdot U_{Fx} + C_2 \cdot U_{Fy} + C_3 \cdot U_{Fz} + C_4 \cdot U_{Mx} + C_5 \cdot U_{My} + C_6 \cdot U_{Mz} \\ M_x = D_1 \cdot U_{Fx} + D_2 \cdot U_{Fy} + D_3 \cdot U_{Fz} + D_4 \cdot U_{Mx} + D_5 \cdot U_{My} + D_6 \cdot U_{Mz} \\ M_y = E_1 \cdot U_{Fx} + E_2 \cdot U_{Fy} + E_3 \cdot U_{Fz} + E_4 \cdot U_{Mx} + E_5 \cdot U_{My} + E_6 \cdot U_{Mz} \\ M_z = F_1 \cdot U_{Fx} + F_2 \cdot U_{Fy} + F_3 \cdot U_{Fz} + F_4 \cdot U_{Mx} + F_5 \cdot U_{My} + F_6 \cdot U_{Mz} \end{cases} \tag{2}
$$

Conclusions after preliminary tests of the developed measurement application
The conducted laboratory tests carried out allowed to check the developed program and correct minor application errors. They confirmed the possibilities of building a graphical measurement environment to support complex measurement systems equipped with various measuring devices. In the described version, the program allows support for Hottinger Baldwin Messtechnik amplifiers (MGCPlus, Spider8, QuantumX), several multifunction measuring cards (DAQ) from National Instruments and Tektronix digital multimeters and oscilloscopes. Unfortunately, the anticipated constraints resulting from the capabilities of measuring devices were also confirmed, especially when using communication interfaces with significantly different data transmission rates. During testing at high sampling frequencies, a larger number of measurement signals proved to be a significant limitation for the computer used during measurements as a control unit. The tests carried out in the Windows environment have shown that in order to increase the requirements for the number of measured quantities, the amount of data that must be processed (all online calculations: taking into account the characteristics and virtual channels) visualized and stored in the system buffer for possible archiving. All this requires a computer with a powerful processor and lots of memory. The authors also noted the need to extend the program's capabilities to cache data in the computer's mass memory (e.g. a high-speed hard drive) so that the buffer size in the memory would not limit the amount of data to be stored during longer tests.

Experimental Mechanics of Solids   Materials Research Forum LLC
Materials Research Proceedings **12** (2019) 117-123   https://doi.org/10.21741/9781644900215-17

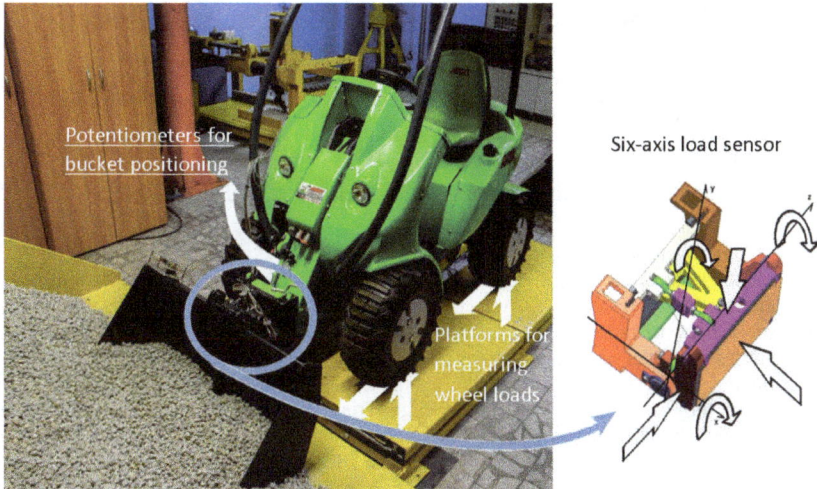

Fig. 6. A new-generation test stand with a measuring system to identify the loads of the work tool (bucket) and wheels of the undercarriage and the position of bucket [6]

Summary

The presented computer measuring program created in the LabVIEW environment enables the implementation of complex research programs with the visualization and registration of all the measured quantities. The open structure of the system allows for a relatively simple expansion with further elements: e.g. for automatic control of the executive elements, support for other measuring devices. Particular attention requires the expansion of the equation editor to define virtual channels and the use of mass storage as a buffer with more data.

References

[1] National Instruments (2017). LabVIEW, http://www.ni.com/LabVIEW.

[2] A. Suliman, Applications of Virtual Modules in Numerical Analysis, Open Access Library Journal, 3, (2016) 1-8.

[3] H. Mageed and A. El-Rifaie, "Electrical Metrology Applications of LabVIEW Software," Journal of Software Engineering and Applications, Vol. 6 No. 3, 2013, pp. 113-120. https://doi.org/10.4236/jsea.2013.63015

[4] M. Alia, T. Younes and M. Zalata, Development of Equivalent Virtual Instruments to PLC Functions and Networks Journal of Software Engineering and Applications, Vol. 4 No. 3, 2011, pp. 172-180. https://doi.org/10.4236/jsea.2011.43019

[5] R. Bitter, T. Mohiuddin and M. Nawrocki, LabVIEW Advance Programming Techniques, Second Edition, CRC Press, Boca Raton, 2006. https://doi.org/10.1201/9780849333255

[6] R. Czabanowski, Ł. Leśniak, A. Łabuda, The computer measuring system of the stand for testing the load process with a bucket of a one-bucket loader, AUTOBUSY, No. 12, 2016 (in Polish).

Experimental Mechanics of Solids   Materials Research Forum LLC
Materials Research Proceedings **12** (2019) 124-130   https://doi.org/10.21741/9781644900215-18

The Development of Fatigue Cracks in Metals

Dariusz Rozumek

Opole University of Technology, Mikolajczyka 5, 45-271 Opole, Poland

d.rozumek@po.opole.pl

Keywords: Fatigue Crack Growth, Microstructure, Load, Mixed Mode

Abstract. The work presents the development of fatigue cracks in metals (review article). In particular, the reasons for the growth of fatigue cracks, models of the development of cracks, cracks initiation and propagation for various modes as well as effects of the cracks growth were presented. Fatigue of materials, especially the formation of fatigue cracks and their growth, belong to the important problems of solid mechanics. Designers and constructors of machines and industrial devices are focusing their attention on problems concerning durability and reliability of these devices. Therefore, the engineering materials that are used to construct devices should have the best properties, because choosing the right material affects, to a large extent, the durability of construction.

Introduction

Fatigue of the material occurs under the influence of variable loads (stresses) over time, which are typical in various machine systems [1,2,3]. It is a very important phenomenon as it is the main culprit behind about 90% of damaged machine parts [4]. This process occurs through nucleation and crack growth.

Mechanics of fracture is a relatively young science branch. Its development took place during the Second World War. One of the reasons for interest in this phenomenon was the failure of the "Liberty" type ships. The development of fatigue cracks in those ships concerned brittle fracture, which occurred due to existing material defects in the welding process. Brittle fracture is very dangerous due to the very rapid cracks growth and cause the greatest material losses compared to other types of fractures (ductile, mixed) [4]. Three parameters are used to describe fatigue crack growth: stress, displacement and energy. The stress parameter describes the stress state near the crack tip in brittle materials with an error of 5 - 20% compared to elastic-plastic materials for the stress range $0.4\sigma_y < \sigma < 0.7\sigma_y$ [4]. The displacement parameter is described with the use of the crack tip opening displacement δ. Since the crack tip opening displacement (CTOD) is a certain measure of strains in the area near the tip, it is also called a "strain parameter". This parameter is applied for the elastic-plastic materials to the yield point σ_y. The energy based parameters use the strain energy density and the J-integral [5].

The aim of the paper is to present the causes of fatigue cracks growth, models for its description, as well as test results and effects caused by cracks on examples of structural damage.

Causes of fatigue cracks

Structure and surface of the material

One of the reasons for the development of fatigue cracks are point and linear crystal defects (dislocations) [6-8]. We can distinguish four types of point defects in the material (see Fig. 1) as: an atom of a foreign element - 1, additional atom of a foreign element - 2, void (lack of atom) - 3, indigenous atom in the wrong place - 4. There are two types of dislocations: edge (presence of an additional crystalline half-plane) and screw (displacement of a part of the crystal). Dislocations are usually mixed. In areas of special dislocation concentration, slip bands may

Experimental Mechanics of Solids
Materials Research Proceedings **12** (2019) 124-130

Materials Research Forum LLC
https://doi.org/10.21741/9781644900215-18

occur. There are plastic strains in the slip bands. They are related to the strains inside the crystals, which may be elastic or plastic. Elastic strains cause tension, which disappears after removing the load. Plastic strains break the atomic bonds, as a result of which the atoms gain new neighbors. Plastic strains do not occur in the whole volume of the crystal, but as a dislocation movement most conveniently oriented relative to the τ_{max}. As a result, followed by a break of only some atomic bonds. Metals can be divided into elastic and plastic based on their mechanical properties.

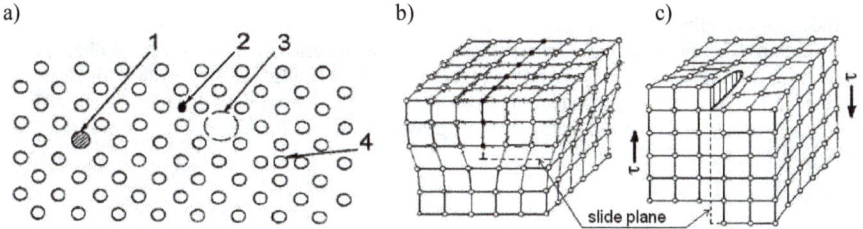

Fig. 1. Types of point defects in the crystal lattice of the metal a), two types of dislocations: b) edge, c) screw [9]

Elastic metals (among others brittle) have a high strength and as a result: (i) have few slip bands, (ii) the initiations of microcracks are in the places of defects, (iii) microcracks are less common than in ductile metals, (iv) the increase in size of microcracks occur in planes perpendicular to the load, and (v) microcracks connect into macrocracks. Plastic (ductile) metals have low strength and as a result one can see that: (i) the number of slip bands increases with the number of cycles up to the saturation level, (ii) the development of plastic strains occurs only in some slip bands, (iii) some of the slip bands are transformed into microcracks inside a crystal structure, (iv) microcracks growth and connect up to the appearance of macroscopically visible cracks (length about 10^{-1} mm), (v) the growth of macrocracks causes damage.

Load and environment

Cracks in the material are caused by occurring stresses during load (Fig. 2a). More precisely, the lack of stresses relaxation at the tip of the crack due to dissipation processes (development of plastic strains), which causes the slip bands in the direction of τ_{max}.

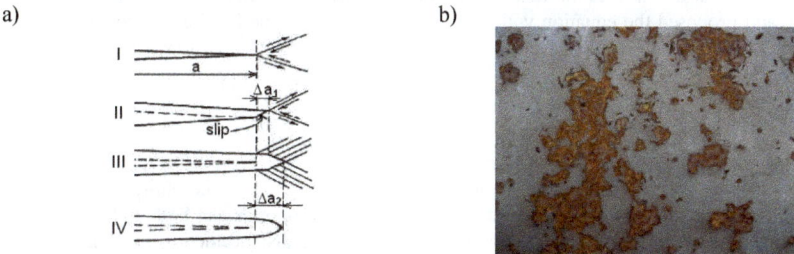

Fig. 2. Mechanism of fatigue cracks growth a), steel surface with defects caused by corrosion b)

Experimental Mechanics of Solids
Materials Research Proceedings **12** (2019) 124-130

Materials Research Forum LLC
https://doi.org/10.21741/9781644900215-18

The effect of this is crack growth by Δa (Fig. 2a). The creation of new slip bands results in a further increases in the crack and blunting (rounding) of its tip. After unloading, the process is repeated until damage occurs. Crack growth is often accelerated by the influence of the environment (eg. corrosion – Fig. 2b) to which the material is exposed. It causes the oxidation of the material and the formation of defects creating cracks, pits and etc. Corrosion acceleration can occur under the influence of moisture, as well as chlorine or sulfur. In addition, the corrosion rate is affected by the temperature and the higher it is, the faster the processes take place. With simultaneous stresses and corrosion, the crack growth is more intense.

Models to describe fatigue cracks

In addition to brittle and ductile fracture, we can divide the cracking based on the Wöhler curve. The fatigue crack growth may be in the low cycle fatigue (LCF) and high cycle fatigue (HCF) range. In the case of LCF, initiation and propagation occurs from the beginning of the process (Fig. 3a), and for HCF, initiation and cracks growth occurs at the end of element life (see Fig. 3b) [10].

a)   b)

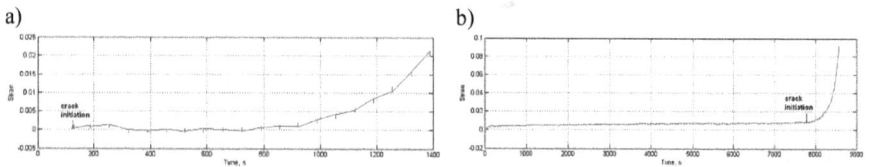

Fig. 3. Crack initiation point for a) LCF, b) HCF

Irwin proposed in 1957 [11] the stress intensity factor (SIF) represented by the variable, K. SIF describes the stress state at a crack tip, is related to the crack growth rate, and is used to establish the failure criteria due to fracture. According to Irwin's proposal, there exist three types of loading: mode I - tensile opening, mode II - in-plane sliding, mode III - tearing or anti-plane shear [5]. SIF was used to create curves of fatigue crack growth rate. One of the first was Paris [12], who proposed for linear elastic fracture mechanics (LEFM) relation

$$da/dN = C(\Delta K)^m \tag{1}$$

where C and m are material constants determined experimentally. The main weakness of the Paris law is a fact that it does not take into account the effect of mean stress, what Forman [13] noticed, and proposed the equation with an additional critical value K_c (fracture toughness) and stress ratio R

$$da/dN = \frac{C(\Delta K)^m}{(1-R)K_c - \Delta K} \tag{2}$$

The next modification of Eq. (1) was done by Elber taking into account the crack closing and opening. Priddle was one of the first who proposed a description of the entire crack kinetics curve. The next authors were McEvily and Yarema, whose proposals can be found in the paper [14]. Depending on the applied stress, strain (crack-tip opening displacement - CTOD), or energy approach, the fatigue crack growth rate can be represented as a function of one of the mentioned parameters. For example in Ref. [14], the author proposed an equation for the description of the

Experimental Mechanics of Solids   Materials Research Forum LLC
Materials Research Proceedings **12** (2019) 124-130   https://doi.org/10.21741/9781644900215-18

total curve of crack growth kinetics versus the energy parameter range ΔJ, i.e. from the range of the threshold value ΔJ_{th} to the critical value of the parameter J_{Ic}.

$$da/dN = B \left[\frac{\Delta J - \Delta J_{th}}{(1-R)^2 J_{Ic} - J_{max}} \right]^n,$$ (3)

where B and n are experimental coefficients, and J_{max} is the maximum value of parameter J.

The development of fatigue cracks

The development of fatigue cracks were carried out on the machines shown in Figs 4 and 5. Fig. 4a shows the fatigue test stand MZPK 100 [15] for cruciform specimens (mixed mode I+II), which allows to conduct fatigue experiments under cyclic and random loading, also with a static mean value. In the holed specimen, made of S355 steel, shown in Fig. 4b, the cracks developed for mode I. In the solid specimen shown in Fig. 4c the cracks developed for mixed mode I+II then for mode I. The MZGS-100 stand [16,17] for testing the fatigue crack growth during

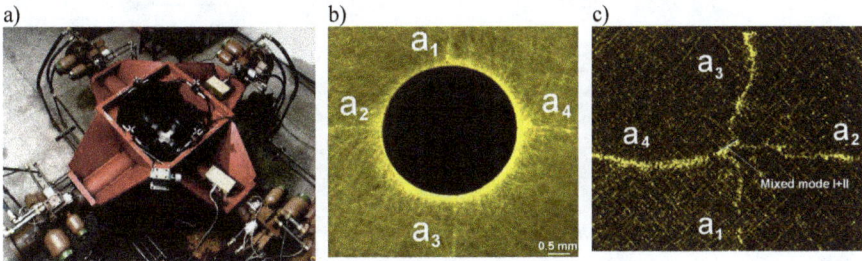

Fig. 4. The MZPK-100 fatigue stand setup a), crack growth in specimens: b) with hole, c) without hole

proportional bending with torsion (mixed mode I+III) and additional static (mean) load (Fig. 5a). The specimen load for combination of bending with torsion (M_{BT}) is shown in Fig. 5b.

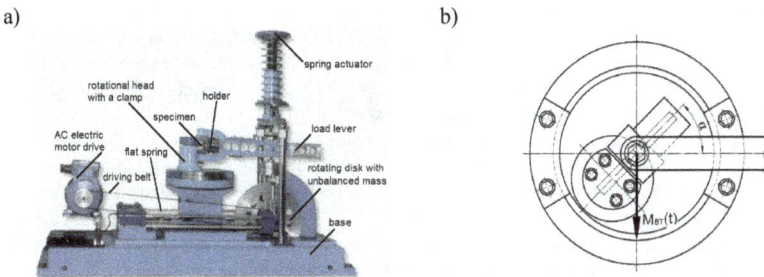

Fig. 5. The MZGS-100 fatigue stand setup a), loading of the specimen b)

Experimental Mechanics of Solids
Materials Research Proceedings 12 (2019) 124-130

Materials Research Forum LLC
https://doi.org/10.21741/9781644900215-18

The length of the crack versus the number of cycles is used to describe the fatigue cracks growth. Depending on the applied load or occurring notches [18], the growth of cracks may have a different lifetime. Fig. 6 presents the fatigue cracks length versus the number of cycles for S355 steel under a different values of the bending moment amplitudes (M_a) and various notches (K_t). From Fig. 6 it appears that an increase of the bending moment and change the notch root radii from blunt to sharp causes a decrease in the fatigue life of the tested specimens.

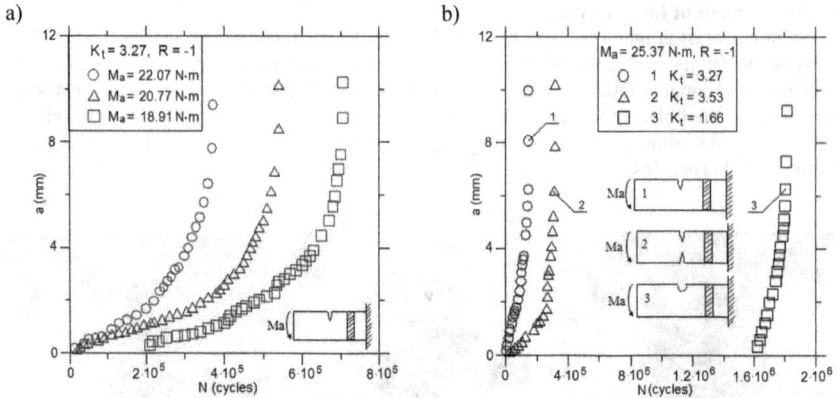

Fig. 6. Fatigue crack lengths versus the number of cycles for different a) loads, b) notches

The effects of the fatigue cracks growth

The growth of fatigue cracks lead to damage or destruction of the object or structure. The effects of fatigue cracks are observed on the examples of various disasters. Damage of Liberty ships is one of the most frequently cited examples of brittle fractures. The need for fast and cheap to build ships during the Second World War resulted in the replacement of sheet metal riveting with welding, which significantly accelerated the production of ships (until 1946, about 4,000 ships were built). In the years 1941-1953, there were 1250 ships damaged by brittle fractures, of which 12 ships broke into two parts. Research on the cause of the damage ships began when one of the ships in a dry dock in the shipyard in Kensington Bay broke in a half. These cracks occurred in places of stress concentration (most often in the middle of the ship), and their development was initiated by welding defects (the used material had very low impact resistance at low temperatures). There are various other examples such as the sinking of the oil platform Alexander Kielland, as well as some plane crashes, train accidents and bridge disasters. The cause of loss of stability of the Kielland oil platform was a fatigue fracture of the reinforcing element to which the hydrophone was welded. In the years 1952-1954 a series of plane crashes with the use of the Comet aircraft resulted in many deaths. Experimental research with variable pressure differences across the entire Comet plane structure showed the appearance of cracks in the aircraft fuselage near the windows and safety exits. In 1980, a plane crashed near the Okęcie airport, in which 77 passengers and 10 crew members were killed. The cause was a fatigue fracture of the engine shaft. In 1998, a passenger train in Eschede (Germany) derailed, in which 101 people were killed and 88 injured. The cause of the derailment of the train was fatigue damage to one of the wagon wheels. Another train derailed in Spain (in 2001) where 5 passengers were injured. The cause was the fatigue fracture of the rail under the train. In 2007, a truss bridge over the Mississippi River collapsed. The reason was the corrosion of metal elements and material fatigue which caused the bridge to lost its flexibility.

Summary

Fatigue microcracks usually appear on the surface or in the surface layer of an object. In only a few cases, e.g. in surface-improved elements, the microcracks can develop on the edge of the core and the hardened layer. Similarly, in elements with galvanic coatings, it is possible to initiate cracks at the boundary of the substrate and coating. The sources of microcracks are also non-metallic inclusions and surface defects of technological origin. Nucleation of fatigue crack occurs at the tip of microcracks. It develops initially in the plane experiencing maximum tangential stresses. The development of fatigue cracks in a corrosive environment is different than in an inert environment. This manifests itself in different reactions compared to material working in a corrosive and neutral environment to changes in the parameters of mechanical loads. The development of fatigue cracks is common in constructions and devices, the effect of which is failure or damage (destruction) of the structure. It causes large financial losses. In many cases, deaths and disability have been caused by the development of fatigue cracks. Statistics show that in the United States loses reach 100 million dollars a year. The main reasons for the development of fatigue cracks (according to British statistics) are: stress concentrators, surface quality, wrong choice of material and not sufficient design experience.

References

[1] S. Suresh, Fatigue of Materials, Cambridge University Press, Cambridge, UK, 1991.

[2] J. Gadomski, P. Pyrzanowski, Experimental investigation of fatigue destruction of CFRP using the electrical resistance change method, Measurement. 87 (2016) 236-245. https://doi.org/10.1016/j.measurement.2016.03.036

[3] L. Sniezek, T. Slezak, K. Grzelak, V. Hutsaylyuk, An experimental investigation of propagation the semi-elliptical surface cracks in an austenitic steel, Int. J. Pressure Vessels and Piping. 144 (2016) 35–44. https://doi.org/10.1016/j.ijpvp.2016.05.006

[4] S. Kocańda, Fatigue Failure of Metals, WNT, Warsaw, 1985.

[5] D. Rozumek, E. Macha, A survey of failure criteria and parameters in mixed-mode fatigue crack growth, Materials Science. 45 (2009) 190-210. https://doi.org/10.1007/s11003-009-9179-2

[6] U. Zerbst, M. Madia, C. Klinger, D. Bettge, Y. Murakam, Defects as a root cause of fatigue failure of metallic components. I: Basic aspects, Eng. Failure Analysis. 97 (2019) 777-792. https://doi.org/10.1016/j.engfailanal.2019.01.055

[7] P. Lukas, L. Kunz, Notch size effect in fatigue, Fatigue Fract. Eng. Mater. Struct. 12 (1989), 175-186. https://doi.org/10.1016/0142-1123(89)90261-2

[8] C. Verdu, J. Adrien, J.Y. Buffiere, Three-dimensional shape of the early stages of fatigue cracks nucleated in nodular cast iron, Mater. Sci. Eng. A. 483–484 (2008), 402-405.

[9] K. Przybyłowicz, J. Przybyłowicz, Repetytorium z materiałoznawstwa, cz. II, Fizyczne podstawy materiałoznawstwa, Politechnika Świętokrzyska, Skrypt nr 279.

[10] D. Rozumek, Z. Marciniak, Fatigue properties of notched specimens made of FeP04 steel, Materials Science. 47 (2012) 462-469. https://doi.org/10.1007/s11003-012-9417-x

Materials Research Forum LLC
https://doi.org/10.21741/9781644900215-18

[11] G.R. Irwin, Analysis of stresses and strains near the end of a crack traversing a plate, Journal of Applied Mechanics. 24 (1957) 361-364.

[12] P.C. Paris, F. Erdogan, A critical analysis of crack propagation laws, J. of Basic Eng., Trans, American Society of Mechanical Engineers. 85 (1960) 528-534.

[13] R.G. Forman, V.E. Kearney, R.M. Engle, Numerical analysis of crack propagation in cyclic-loaded structures, Journal of Basic Eng., ASME. 89 (1967) 459-464. https://doi.org/10.1115/1.3609637

[14] D. Rozumek, Survey of formulas used to describe the fatigue crack growth rate, Materials Science. 49(6) (2014) 723–733. https://doi.org/10.1007/s11003-014-9667-x

[15] D. Rozumek, C.T. Lachowicz, E. Macha, Analytical and numerical evaluation of stress intensity factor along crack paths in the cruciform specimens under out-of-phase cyclic loading, Engineering Fracture Mechanics. 77 (2010) 1808-1821. https://doi.org/10.1016/j.engfracmech.2010.02.027

[16] J. Lewandowski, D. Rozumek, Cracks growth in S355 steel under cyclic bending with fillet welded joint, Theoretical and Applied Fracture Mechanics. 86 (2016) 342-350. https://doi.org/10.1016/j.tafmec.2016.09.003

[17] D. Rozumek, Z. Marciniak, Control system of the fatigue stand for material tests under combined bending with torsion loading and experimental results, Mechanical Systems and Signal Processing. 22 (2008) 1289-1296. https://doi.org/10.1016/j.ymssp.2007.09.009

[18] G. Robak, D. Krzyzak, A. Cichanski, Determining effective length for 40 HM-T steel by use of non-local line method concept, Polish Maritime Research. 25 (2018) 128-136. https://doi.org/10.2478/pomr-2018-0015

Experimental Mechanics of Solids
Materials Research Proceedings **12** (2019) 131-138

Materials Research Forum LLC
https://doi.org/10.21741/9781644900215-19

Steering Kinematics and Turning Resistance Experimental Investigation of Articulated Rigid Body Vehicles

Aleksander Skurjat

Department of Off-road Machine and Vehicle Engineering,

Wroclaw University of Technology,

I. Łukasiewicza 7/9, 50-374 Wrocław,

aleksander.skurjat@pwr.edu.pl

Keywords: Off-Road Vehicle, Articulated Body Steering System, Steering Kinematics

Abstract. Steering kinematics and turning resistance torque depends on geometrical parameters, mass distribution and a type of suspension system of an articulated body steer vehicle with a combination of a ground type. It is important how does a driving system is designed because of its influences on wheels torque while turning. When a vehicle is turning an angle in an articulated joint is rising and the front and rear frame rotate themselves by an angle to the symmetry line when previously situated vehicle forward. In a paper, a method for front and rear frame measuring angles of rotation is presented. There are also results of turning resistances with different mass in articulated wheel loader bucket and two types of ground. Results show, that increasing of a wheeled bucket load influences both steering torque required for turning and changes angles of rotation of the front and rear frame of a vehicle. Steering torque at a first phase is measured to rise to maximum level then decrease even bucket load increase. We can observe that revolution angle of front and rear frame during steering depends strongly on a bucket load, ground type and driving system. Knowledge in rotation angles gives an opportunity to propose a mathematical model for future prediction turning resistance torques.

Introduction

Off-road rigid frames earth working vehicles allows for increased efficiency in construction tasks. There are many difficulties in the proper design of such machines. Most of these vehicles use articulated steering system. Problem with measurements arises when a system is more complex and typically few of coefficients/parameters change their values at the same time. The one of such complex system is articulated hydrostatic steering system of wheeled vehicles with is corresponding with different types of ground. While the vehicle is turning, many parameters like torque arm, oil bulk modulus, a pressure is changing. We have to take into account different ground types too.

A very important disadvantage of off-road vehicles with articulated steering system is low maximum velocity. This is made by snaking phenomena considered as self-path changing due to disadvantages of the steering system. Knowledge how articulated machine steers is crucial in finding equations to calculate and then putting an automatic correction in a path of such vehicle. We can observe (measure) an angle between the front and rear part of a vehicle γ very easily by using most types of angle transducers. The angle measured in the steering joint γ is a sum of two: course angle of front frame γ_p and rear frame γ_t and represents an angle of rotation of the front and rear frame. Course angle, in definition, is an angle measured between the velocity vector of the front or rear frame to the line crossing centres of the front and rear axles. Ratios between γ_p and γ_t depend on many situations: travelling velocity, normal reactions on tires (load in a bucket), vehicles geometry, ground type. Ratio calculation between γ_p and γ_t only from

Experimental Mechanics of Solids   Materials Research Forum LLC
Materials Research Proceedings **12** (2019) 131-138   https://doi.org/10.21741/9781644900215-19

geometrical parameters can produce a significant error and no path correction algorithm can be introduced.

In the article, the author presents an experimental method for estimating the rate between the front and rear course angle (γ_p and γ_t) with correlation to the steering angle (γ). The measurement system consists of the pressure sensor in a chamber of steering cylinder, articulation sensor and the Earth magnetic field sensor, which are used to measure the course angle of a front frame (γ_p).

A second problem with achieving high velocities of articulated vehicles is made by a low stiffness of the hydraulic system. In previous research of authors, obtained measurement values of steering stiffness were presented with contact or without contact tires to the ground. With the use of torque generation device between the front and rear frames M_s, steering stiffness is calculated. By the use of the presented measuring system, the author shows forces between ground and tires in function of articulated angle (turning resistance forces) [4]. The results are obtained by the use of steering system designed by the manufacturer. The aim of measurement is to answer, how load and velocity affect turning resistance in universal equations.

Steering kinematics of articulated vehicles

Steering kinematics of an articulated vehicle at standstill mostly depends on steering system type and geometry. For the vehicles with steering wheels (Ackerman system), a vehicle steers properly when perpendicular lines to the velocity vector of all wheels insect in one point called a turning point. In this case, the moving direction of the vehicle is known and of course with an assumption of a course angle of wheels. Course angle is a result of centrifugal forces, lateral forces etc.

In the case of articulated vehicles, a situation is not identified. The steering system hydraulic actuator makes a rotation of the front and rear frames each another. It is not clear what is a value of turning angle (γ_p and γ_t) of frames to the ground. A theory says that a sum of revolution angles of frames measured to the ground is giving an angle in a steering joint. In real conditions, we can observe two opposed situations. In the first edge situation only one frame rotates (second is fixed), or in a second case, both of them turn the same angle. The result depends on vehicles and steering system geometry, turning resistances, frame masses, tires rolling resistances torques (with a driving system).

Steering system kinematics and turning resistances experimental investigations were performed in the Department of Off-road Machine and Vehicle Engineering, Wroclaw University of Technology. Many research can be found in this field of an experiment [1, 2]. We can observe that when an articulated vehicle is turning at standstill, the middles of centres of differentials O_p and O_t get closer to each other – Fig. 1.

Experimental investigations show, that when turning resistances of a front and rear frame are equal we can observe a symmetrical trajectory p and t, and the same angles of rotation γ_p and γ_t showed on Fig. 2. The case of inequality of turning resistances results in different γ_p and γ_t and even the situation that one frame is fixed showed on Fig. 3. The source of turning resistances torque considering two wheels on driving axle are: rolling resistance torques, braking/driving torques from a drive, tires normal reactions, deformation of a soil. More details considered turning kinematics, specifically forces between tires and a ground and then needed explanations of unused in the article markings and symbols can be found in [1, 2].

Experimental results are giving an answer how does a steering joint angle γ is divided on γ_p and γ_t angles of frames and how does turning resistances are changed to due increasing wheel loaders bucked mass and type of ground.

Experimental Mechanics of Solids   Materials Research Forum LLC
Materials Research Proceedings **12** (2019) 131-138   https://doi.org/10.21741/9781644900215-19

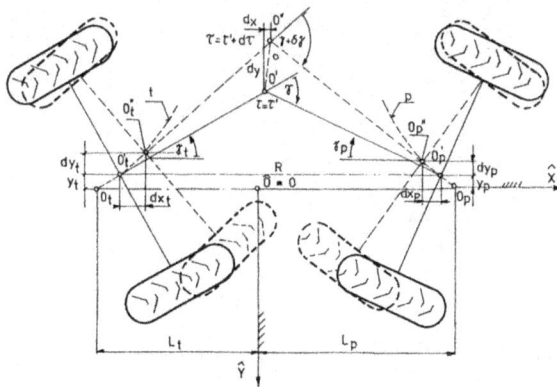

Fig. 1. The turning position of an articulated vehicle and shift trajectories p, t, o [1]

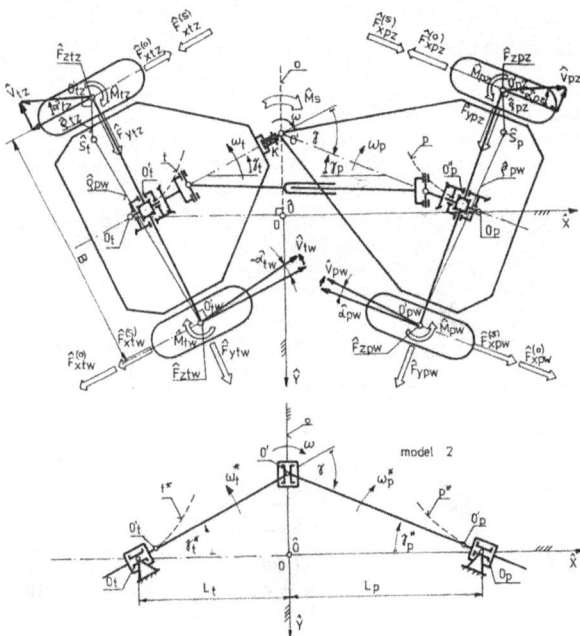

Fig. 2. Steering kinematics while symmetrical turning resistances of frames [1]

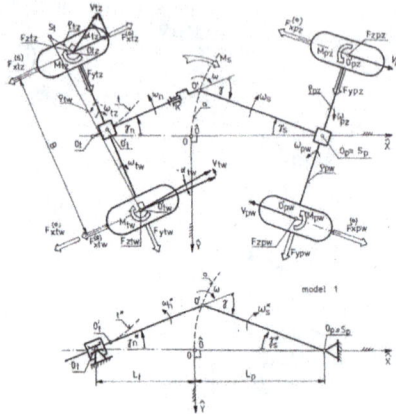

Fig. 3. Steering kinematics while unsymmetrical turning resistances of frames [1]

Identify of an exact mechanism of dividing an angle γ to γ_p and γ_t allows for preparing a control system for traction control [3] and mitigate an effect of articulated vehicles snaking phenomena [4].

Test method and measuring stand

The experiment was conducted on mini wheel loader L052 Fadroma, overall mass m=1470 kg. The vehicle drives all 4 wheels by using a hydrostatic drive. The driving system consists of a hydraulic pump, valves and two hydraulic motors. Each hydraulic motor drives a differential mechanism without a lock to drive wheels. Basic mass and geometry parameters are shown in Fig. 4.

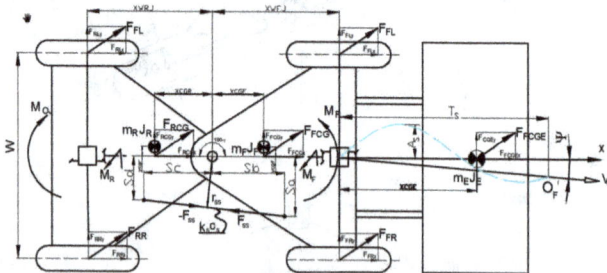

Fig. 4. The basic geometry and mass parameters of mini wheel loader L052, X_{cgr}=0,8; Y_{cgr}=0,2; X_{cgf}=0,7; Y_{cgf}=0,1; X_{wrf}=1,3m; X_{wfj}=1,4; m_r=780kg; m_f=690kg; m_{rf}=300kg; m_{lf}=390kg; m_{rr}=430kg; m_{rl}=350kg

A measuring system consists of 3-axis the Earth magnetic field sensor HMC5883L and 3-axis acceleration sensor ADXL345. Modules were connected to microcontroller system by I2C protocol then the data send by RS232 protocol to Matlab/Simulink software. During the measurements of turning angles of frames by the use of the Earth field magnetic sensor, there is a

Experimental Mechanics of Solids   Materials Research Forum LLC
Materials Research Proceedings **12** (2019) 131-138   https://doi.org/10.21741/9781644900215-19

need to compensate incline of a module and an acceleration sensor was used to compensate results. Measurement uncertainty is estimated to ±0,5° but it strongly depends on environmental the experiment takes place. Author believes that such accuracy is sufficient. In the Matlab/Simulink compensation model correction were calculated by following equations:

$$tg(roll) = \frac{A_y}{A_z} \tag{1}$$

$$tg(pitch) = \frac{-A_x}{A_y \sin(roll) + A_z \cos(roll)} \tag{2}$$

$$tg(yaw) = \frac{(M_z - K_z)\sin(roll) - (M_y - K_y)\cos(roll)}{(M_x - K_x)\cos(pitch) + (M_y - K_y)\sin(pitch)\sin(roll) + (M_z - K_z)\sin(pitch)\cos(roll)} \tag{3}$$

where: A_x, A_y, A_z – measured acceleration, M_x, M_y, M_z- measured angle orientation,
          K_x, K_y, K_z – the Earth magnetic field sensor calibration coefficient,
          roll, pitch, yaw – calculated correction angle.

In the literature [6], [7], [8] a similar compensation method was used and the measuring method was performed with good precision.

Experiments were conducted on two different friction coefficient of the ground. During the first set of measurements friction coefficient was about $\mu=0.62$ for all wheels and next, only the front wheels were situated on $\mu=0.25$. During experiment bucket load was changing form $m_E=0$ to 300 kg. Near right front wheel on front axle symmetry line, a measuring module was mounted. A potentiometer transducer situated between the front and rear frame measures articulation angle. The pressure sensor in a hydraulic cylinder was measuring turning resistance torque. Experimental result while turning with 300 kg load in a bucket and $\mu=0.62$ ground friction coefficient is shown in Fig. 5.

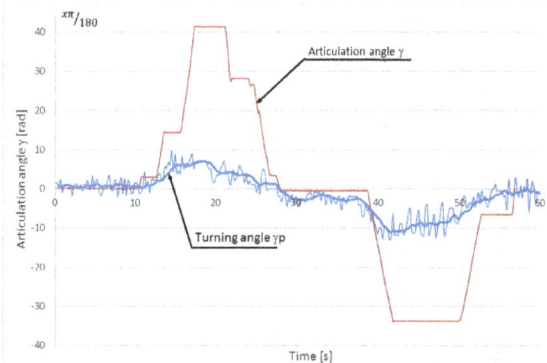

Fig. 5. Changing of γ_p as a result of a vehicle turning γ on a ground with a friction coefficient of $\mu=0.62$ and bucket mass $m_E=300$ kg

Experimental Mechanics of Solids   Materials Research Forum LLC
Materials Research Proceedings **12** (2019) 131-138   https://doi.org/10.21741/9781644900215-19

The experiment results showed higher values for a turning angle γ_p for clockwise rotation to turning the opposite direction for friction coefficient of $\mu=0.62$. We can observe that the ratio γ/γ_p for clockwise rotation obtains $\gamma/\gamma_p \approx 3$ when anticlockwise turning coefficient gets $\gamma/\gamma_p \approx 5$. This is made probably by unsymmetrical normal reaction of the wheels on the right and left side per driving axle. This result in different rolling resistant torque of each wheel.

For ground friction coefficient of $\mu=0.25$, an asymmetry diminishes and the ratio is obtaining values $\gamma/\gamma_p \approx 4$ in both turning directions. This is because a side slip of front wheels makes longitudinal and lateral forces in tire-ground contact becomes smaller and in results, lower braking torques were acting on wheels.

Results obtained during the experiment for different load and friction coefficient of ground have a similar plot. Maximum values of γ_p and γ are presented in Fig. 6.

Results showing how bucket load affects turning resistance force F_s are presented in Fig. 7.

Experimental results showed that on the ground with friction coefficient $\mu=0.62$ turning resistance is rising with bucket mass. This is because of uneven normal reactions of front $m_f= 690kg$ and rear axle $m_r=780kg$. We can observe that turning resistance rises fast until normal reactions on a front and rear wheels gets equal values. Beyond this point, the rear frame is going up because of a high load in a bucket. This result in a situation, that normal reaction on rear wheels becomes smaller and of course, they are losing the capability of putting high lateral and longitudinal forces. This results in obtaining smaller turning resistance forces with a high load in a bucket. Lowering a friction coefficient $\mu=0.25$ is giving similar results but with smaller resistance forces.

Mass m_E [kg]	Coef. μ [-]	Angle γ [deg]	Angle γ_p [deg]	Sum γ [deg]	Sum γ_p [deg]
0	0.62	-34	-11.2		
0	0.62	40	8	74	19.2
100	0.62	-32	-7		
100	0.62	41	12	73	19
200	0.62	-32	-10.5		
200	0.62	41	7.25	73	17.75
300	0.62	-34	-11		
300	0.62	41	7.9	75	18.9
0	0.25	-32	-9		
0	0.25	38	10	70	19
100	0.25	-32	-8		
100	0.25	40	12	72	20
200	0.25	-35	-8		
200	0.25	42	5 (?)	77	13
300	0.25	-30	-10		
300	0.25	40	12	70	22

Fig. 6. Experimental investigation results for different mass and ground friction coefficient

Mass m_E [kg]	$\mu(1)$ [-]	$F_s(1)$ [N]	$\mu(2)$ [-]	$F_s(2)$ [N]
0	0.62	3780	0.25	4390
100	0.62	4300	0.25	4400
200	0.62	4970	0.25	4600
300	0.62	5427	0.25	4573

Fig. 7. Turning resistance force with different bucket mass and ground friction coefficient

Conclusions

Presented measuring method can be used to measure turning angles of vehicles frames. Measuring accuracy depends on environmental conditions because the Earth magnetic field sensor is used. Using the acceleration sensor for the Earth magnetic field sensor tilt module compensation works effectively and increases the precision of measuring device. Estimated error is about ±0,7° in walls of the laboratory. The measurements were performed at a standstill and no additional acceleration than a tilting sensor module was detected. A dynamic impact can influence temporary significant error in angle orientation measurement because of incorrect calculation of a tilt correction value.

The results showed that unsymmetrical normal reactions of wheels, mass in the bucket, friction coefficient have a significant effect on the turning angle of frames and turning resistance torque. At a high tires friction coefficient, an asymmetry rises because of higher resistant torques at a machine side. Lower friction coefficient allows for easier rolling of tires and obtaining symmetry of γ_p and γ_t while turning. Higher the bucket mass at a first phase, increases steering torque but at a second phase it decreases. This is because the overweighting machine over the front tires and lowering the normal reactions at rear tires as a result.

Adding of a second measuring point on a rear frame allows for measuring trajectory, measuring with higher precision γ_p and γ_t, to propose snaking phenomena control systems and driver assist systems.

References

[1] P. Dudzinski, Problems of turning process in articulated terrain vehicles, Journal of Terramechanics, ISTVS, vol 19, No. 4, p.246-256, 1983. https://doi.org/10.1016/0022-4898(83)90030-7

[2] P. Dudzinski, Lenksysteme für Nutzfahrzeuge, Springer-Verlag, 2006

[3] A. Skurjat, A. Kosiara, Directional stability control of body steer wheeled articulated vehicles. W: Dynamical Systems in Applications, Łódź, Poland, December 11-14, 2017, Springer, cop. 2018. s. 363-371. https://doi.org/10.1007/978-3-319-96601-4_33

[4] P. Dudziński, A. Skurjat, Research on the influence of geometric parameters on the phenomenon of snaking of articulated vehicles. Zeszyty Naukowe - Wyższa Szkoła Oficerska Wojsk Lądowych im. gen. T. Kościuszki. 2017, t. 186, vol. 49, nr 4. https://doi.org/10.15804/athena.2016.50.04

[5] P. Dudziński, G. Hapel, A. Skurjat, Patent. Polska, nr 216468. Sposób i urządzenie do automatycznego adaptacyjnego sterowania pojazdem nr 388851, 2014

Experimental Mechanics of Solids   Materials Research Forum LLC
Materials Research Proceedings **12** (2019) 131-138   https://doi.org/10.21741/9781644900215-19

[6] S. Madgwick, Automated calibration of an accelerometers, magnetometers and gyroscopes - A feasibility study, 2010

[7] Pablo Esteban Quiroga Garcia Wenjie Li, On Indoor Positioning for Mobile Devices, Department of Signals and Systems, Chalmers University of Technology, Goteborg, Sverige, 2011

[8] C. Treffers, L. Wietmarschen, Position and orientation determination of a probe with use of the IMU MPU9250 and a ATmega328 microcontroller, TuDelft, 2016

Experimental Mechanics of Solids
Materials Research Proceedings **12** (2019) 139-145

Materials Research Forum LLC
https://doi.org/10.21741/9781644900215-20

Comparative Analysis of Methods for Determining Cyclic Properties of Metals

Stanisław Mroziński[1, a*], Zbigniew Lis[1, b]

[1]University of Technology and Life Sciences, Faculty of Mechanical Engineering, Kaliskiego 7, 85-789, Bydgoszcz, Poland

[a]stanislaw.mrozinski@utp.edu.pl, [b]zbigniew.lis@utp.edu.pl

Keywords: Low-Cycle Properties; Fatigue Life; P91 Steel X10CRMOVNB9-1

Abstract. The study compares several methods for determining material data required for fatigue life calculation. The test methods were compared at ambient temperature (T=20°C) and increased temperature (T=600°C). The results show the applicability of simplified methods for determining material data.

Introduction

A series of fatigue tests is usually required to determine low-cycle fatigue properties of metals. Technical details of the tests and the results are included in [1,2], however, due to low loading frequencies and high cost of equipment, those tests are both costly and time consuming.

A simplified method for determining static properties is a stepwise increasing load testing procedure - Lo-Hi [3,4]. The procedure involves loading a single specimen with a programmable load, where the load is increased after a certain number of cycles (Fig. 1). Despite many disadvantages [5], the simplified method for determining cyclic properties is commonly used in practice. Its main advantages are simplicity and quick availability of the results. Stepwise increasing load testing is usually carried out under constant total strain (ε_{ac}=const) or constant plastic strain (ε_{ap}=const). An advantage of the tests carried out in those conditions is lack of material flow which can be observed in tests under σ_a=const. The test conditions used depend on the intended use of the material data [6].

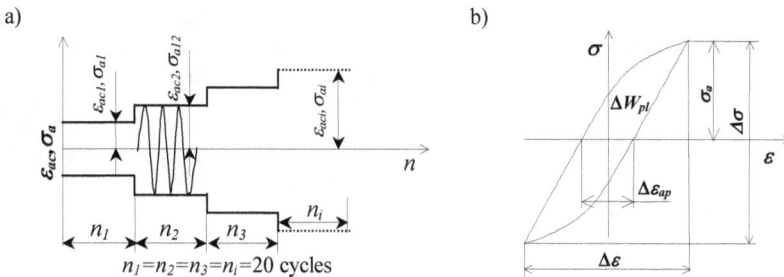

Fig. 1. Increasing load test under ε_{ac}=const and σ_a=const (a), parameters of the hysteresis loop (b).

The study aimed to determine the effects of the method used and the test conditions for P91 steel on material data used for fatigue life calculations. The scope includes low-cycle fatigue tests at $T=20°C$ and $T=600°C$ under two loading conditions (ε_{ac}=const and σ_a=const). Standard analytical models proposed in [1,2] were used in the comparative analysis of the test results.

Analytical characteristics of the hysteresis loop

The basic characteristic describing cyclic properties is a relationship between stress and strain - a cyclic stress-strain curve. The curve can be obtained from experimental data by connecting the peaks of stabilized hysteresis loops obtained at different strain levels. Different models are used to analytically characterize the curve, including a commonly used Ramberg - Osgood model [7]:

$$\frac{\Delta\varepsilon}{2} = \frac{\Delta\sigma}{2E} + \left(\frac{\Delta\sigma}{2K'}\right)^{\frac{1}{n'}} \tag{1}$$

n' and K' parameters are determined in fatigue tests [1, 2], whereas the material constant E is determined in static tensile test. The method for determining n' and K' parameters assumes that plastic strain amplitude ε_{ap} is an exponential function of stress amplitude σ_a and can be expressed as:

$$\varepsilon_{ap} = \left(\frac{\sigma_a}{K'}\right)^{\frac{1}{n'}} \tag{2}$$

The literature also includes other characteristics of the cyclic stress-strain curve using single or two-parameter models [8]. The study focuses on n' and K' parameters of the model expressed as (1). The equation of an ascending branch of the hysteresis loop can be obtained by multiplying the relationship (1) by 2:

$$\Delta\varepsilon = \frac{\Delta\sigma}{E} + 2\left(\frac{\Delta\sigma}{2K'}\right)^{\frac{1}{n'}} \tag{3}$$

The descending branches of the hysteresis loop can be obtained using equation (3) by transforming the coordinate system to the peak of the hysteresis loop. The methods are discussed in detail in the literature [9]. The equations are used to characterize the hysteresis loops for materials showing Masing behaviour [10]. For materials showing non-Masing behaviour, a special plot formed by the upper and lower branches of the hysteresis loop [9] is used. When calculating the fatigue life, the hysteresis loop energy ΔW_{pl} is calculated after the tests under controlled stress (σ_a=const) or controlled strain (ε_a=const, ε_{ap}=const). Using known n' and K' parameters, the energy ΔW_{pl} can be calculated for any $\Delta\sigma$ and $\Delta\varepsilon$ level:

$$\Delta W_{pl} = \Delta\sigma \cdot \Delta\varepsilon - \frac{\Delta\sigma^2}{E} - 4\frac{\Delta\sigma^{\frac{1}{n'}+1}}{\left(2K'\right)^{\frac{1}{n'}} \cdot \left(\frac{1}{n'}+1\right)} \tag{4}$$

Test method

The fatigue test specimens were made with P91 steel (X10CRMOVNB9-1). The fatigue tests were carried out at two temperatures $T=20°C$ and $T=600°C$ under controlled strain ε_{ac}=const and stress σ_a=const. Constant-amplitude fatigue tests under controlled strain ε_{ac}=const were carried out at five strain levels ε_{ac} (0.25; 0.3; 0.35; 0.5; 0.6%). Constant-amplitude fatigue tests under σ_a=const were also carried out at five stress levels. The stress levels σ_a were determined based on the analysis of low-cycle fatigue test results under ε_{ac}=const. Stepwise increasing load test under controlled strain (ε_{ac}-Lo-Hi) started at 0.1%. The load was increased by 0.05% every 20 cycles. Under controlled stress (σ_a-Lo-Hi), the levels were determined based on the analysis of the results of stepwise increasing load test under ε_{ac}-Lo-Hi. Fig. 2 shows the method for determining stress (σ_a).

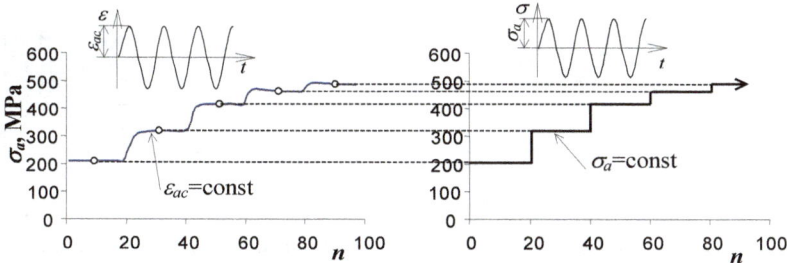

Fig. 2. Determining stress levels under σ_a-Lo-Hi.

Test results

The fatigue tests show changes in the hysteresis loop parameters as a function of the number of load cycles. Fig. 3 shows changes in plastic strain ε_{ap} during constant-amplitude and stepwise increasing load tests as a function of fatigue life n/N. Under σ_a=const, the range of changes in plastic strain ε_{ap} is significantly higher than the range ε_{ap} observed under ε_{ac}=const. It applies to both constant-amplitude tests (Fig. 3a) and stepwise increasing load tests (Fig. 3b).

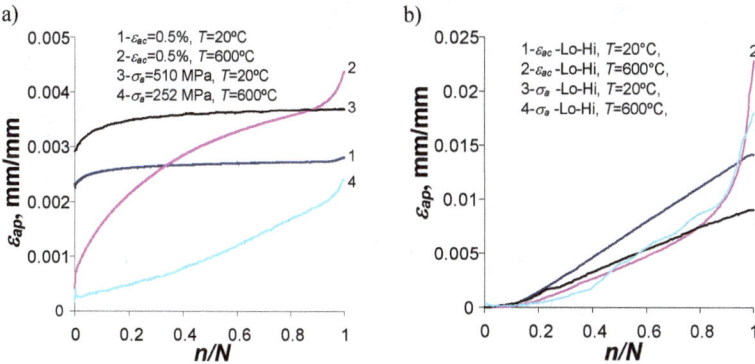

Fig. 3. Changes in ε_{ap}: a) constant-amplitude tests; b) stepwise increasing load tests.

n' and K' material parameters

Cyclic stress-strain curve parameters (equation 2, n' and K') were determined for 0.5 n/N. For tests with stepwise increasing load it corresponds to half of the number of cycles for a single stage. σ_a and ε_{ap} were approximated by regression lines (2) and shown in Fig. 4.

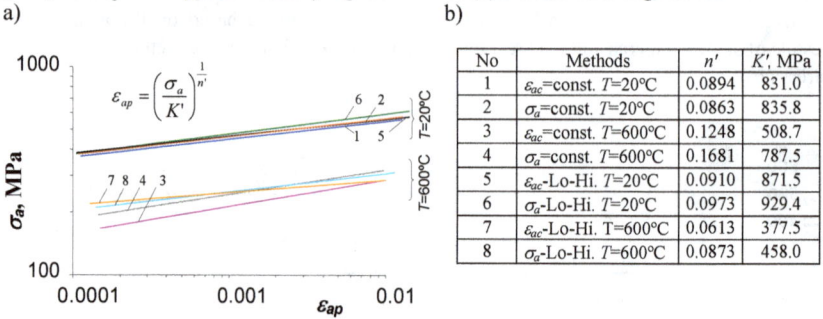

a)

b)

No	Methods	n'	K', MPa
1	ε_{ac}=const. T=20°C	0.0894	831.0
2	σ_a=const. T=20°C	0.0863	835.8
3	ε_{ac}=const. T=600°C	0.1248	508.7
4	σ_a=const. T=600°C	0.1681	787.5
5	ε_{ac}-Lo-Hi. T=20°C	0.0910	871.5
6	σ_a-Lo-Hi. T=20°C	0.0973	929.4
7	ε_{ac}-Lo-Hi. T=600°C	0.0613	377.5
8	σ_a-Lo-Hi. T=600°C	0.0873	458.0

Fig. 4. Cyclic stress-strain curves a) graphical representation; b) n' and K' parameters.

The tests showed that at T=20°C and T=600°C, the relationship between strain ε_{ap} and stress σ_a (2) to a lesser degree depends on the loading conditions. It can be verified by similar n' and K' parameter values. The statistical analysis included a parallelism test and an absolute term test of the regression lines (2) for different loading conditions. The tests show that there are no grounds to reject the hypothesis about the equality of slopes n' and absolute term K' of the analysed regression lines. It verifies the applicability of a simplified test to determine cyclic properties.

Experimental verification of the analytical characteristics of the hysteresis loop

Figs 5 and 6 shows the example hysteresis loops for two strain levels (ε_{ac}=0.5%, 0.6%) at T=20°C and T=600°C obtained from tests and calculations. An equation of the descending and ascending branches of the hysteresis loop was obtained by transforming equation (1).

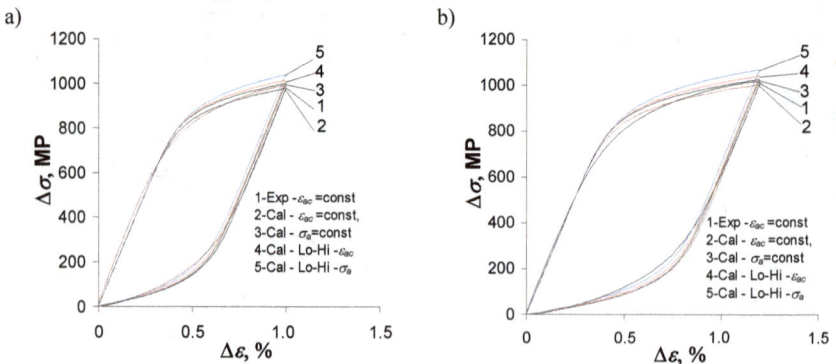

a)

b)

Fig. 5. Hysteresis loops obtained from tests and calculations at T=20°C a) ε_{ac}=0.5%; b) ε_{ac}=0.6%.

Experimental Mechanics of Solids

Materials Research Proceedings **12** (2019) 139-145

Materials Research Forum LLC

https://doi.org/10.21741/9781644900215-20

a)

b)

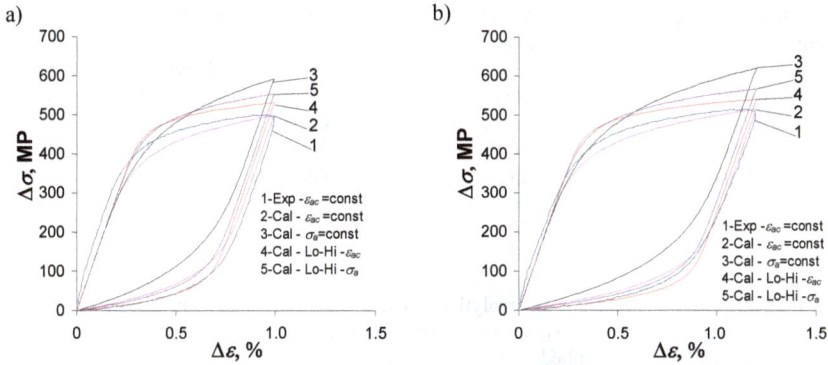

Fig. 6. Hysteresis loops obtained from tests and calculations at T=600°C a) ε_{ac}=0.5%; b)
ε_{ac}=0.6%.

Observed differences in equation (2) parameters (Fig. 5 and 6) resulted in different hysteresis loops. The difference in shape of the hysteresis loops is affected both by temperature and loading conditions during determination of n' and K' parameters.

A comparative analysis of the hysteresis loops obtained from calculations and tests at T=20°C and T=600°C shows larger differences between loops obtained from calculations and tests at T=600°C. Larger differences between the loop obtained from tests and the loop obtained from calculations can be explained by larger changes in loop parameters as a function of the number of loading cycles (Fig. 3).

Coefficient $\delta = \Delta W_{pl(exp)} / \Delta W_{pl(cal)}$ was calculated to qualitatively compare methods for determining n' and K' parameters, where: $\Delta W_{pl(exp)}$ is the hysteresis loop energy obtained from tests and $\Delta W_{pl(cal)}$ - is the energy obtained from calculations. Energy $\Delta W_{pl(cal)}$ was calculated from (4) and the material data determined under different loading conditions. The hysteresis loop energy was calculated for five different strain levels during constant-amplitude tests under ε_{ac}=const. Fig. 7 shows the δ coefficient values.

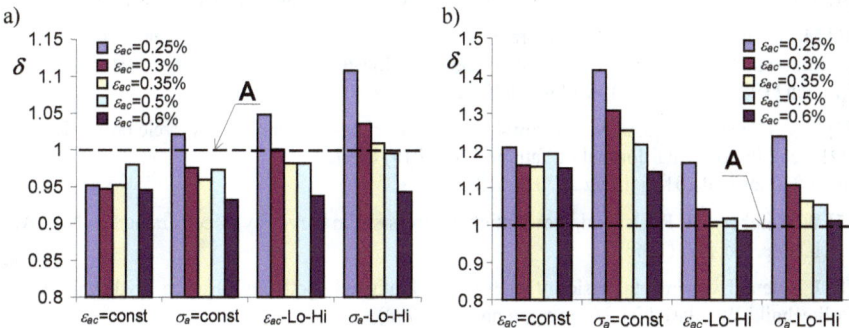

a)

b)

Fig. 7. Coefficient δ: a) T= 20°C, b) T=600°C.

Experimental Mechanics of Solids   Materials Research Forum LLC
Materials Research Proceedings **12** (2019) 139-145   https://doi.org/10.21741/9781644900215-20

Coefficients δ equal to one (line A) indicate that the energy $\Delta W_{pl(exp)}$ obtained from tests at a certain ε_{ac}=const level is equal to the energy $\Delta W_{pl(cal)}$ obtained from calculations and shows the consistency of calculation and test results. The analysis of coefficient δ shows that the test method providing very good representation of the test results cannot be clearly identified. A comparative analysis of δ coefficient at ambient and increased temperature shows that the effectiveness of analytical models describing the hysteresis loop is higher at T=20°C than at T=600°C. δ coefficient values show that at T=20°C, for a majority of strain levels, $\Delta W_{pl(cal)}>\Delta W_{pl(exp)}$, whereas at T=600°C, the opposite is true $\Delta W_{pl(cal)}<\Delta W_{pl(exp)}$.

Summary
The Ramberg-Osgood model - an analytical model describing cyclic properties - includes parameters determined during cyclic loading (n' and K'). A simplified method yields similar values for all parameters controlled throughout the test.

High similarity of the test results obtained using the simplified method (Lo-Hi tests) and the standard method (constant-amplitude tests) shows the applicability of the results obtained using the simplified method for initial calculations and choosing structural features if comprehensive fatigue characteristics are not available.

The material data determined using the simplified method can be used repeatedly when calculating fatigue life [11]. It means, that even the slightest differences in ΔW_{pl} energy obtained from calculations and tests may lead to accumulation of small differences and thus to significant differences between the results obtained from calculations and tests.

This publication is financed by the National Science Centre as part of the research project no. 2017/25/B/ST8/02256

References

[1] PN-84/H-04334 Low-cycle fatigue test for metals.

[2] ASTM E606-92: Standard Practice for Strain – Controlled Fatigue Testing.

[3] S. Mroziński, Analiza porównawcza dwóch metod wyznaczania własności cyklicznych metali, Przegląd Mechaniczny Nr 4 (2004) 30-36.

[4] S. Kocańda, A. Kocańda, Low-cycle fatigue strength of metals, PWN Warsaw (1989).

[5] M. Bayerlein, H. Christ, H. Mughrabi, A critical evaluation of the incremental step test, II International Conference on Low Cycle Fatigue and Elasto-Plastic Behaviour of Materials, Munich (1987) 149-153. https://doi.org/10.1007/978-94-009-3459-7_22

[6] S. Mroziński, H. Egner, M. Piotrowski, Effects of fatigue testing on low-cycle properties of P91 steel, International Journal of Fatigue 120 (2019) 65–72. https://doi.org/10.1016/j.ijfatigue.2018.11.001

[7] W. Ramberg, W.R. Osgood, Description of stress-strain curves by three parameters, NACA, Tech.Note, No 402, (1943).

[8] J. Kaleta, Experimental basics of energy-based fatigue hypotheses, Oficyna Wydawnicza Politechniki Wrocławskiej, Monograph no. 24, (1998).

Experimental Mechanics of Solids   Materials Research Forum LLC
Materials Research Proceedings **12** (2019) 139-145   https://doi.org/10.21741/9781644900215-20

[9] F. Ellyin, D. Kujawski, Plastic strain energy in fatigue failure, J. Pressure Vessel Technology, Trans. ASME 106 (1984) 342-347. https://doi.org/10.1115/1.3264362

[10] Z. Zhang, Z. Hu, S. Schmauder, M. Mlikota K. Fan, Low-Cycle Fatigue Properties of P92 Ferritic-Martensitic Steel at Elevated Temperature, Journal of Materials Engineering and Performance Volume 25 (2016) 1650-1662. https://doi.org/10.1007/s11665-016-1977-8

[11] D.G. Pavlou, The theory of the S-N fatigue damage envelope: generalization of linear, double linear, and non-linear fatigue damage models. International Journal of Fatigue, 110 (2018) 204-214. https://doi.org/10.1016/j.ijfatigue.2018.01.023

Experimental Mechanics of Solids

Materials Research Proceedings **12** (2019) 146-154

Materials Research Forum LLC

https://doi.org/10.21741/9781644900215-21

Fatigue Damage Analysis of Offshore Structures using Hot-Spot Stress and Notch Strain Approaches

António Mourão[1, a*], José A.F.O. Correia[1, b*], José M. Castro[1], Miguel Correia[2], Grzegorz Lesiuk[3], Nicholas Fantuzzi[4], Abílio M.P. De Jesus[1], Rui A.B. Calcada[1]

[1] Faculty of Engineering, University of Porto, Rua Dr. Roberto Frias, 4200-465 Porto, Portugal

[2] Force Technology Norway AS, Hvalstad, Norway

[3] Wroclaw University of Science and Technology, Faculty of Mechanical Engineering, Department of Mechanics, Materials Science and Engineering, Poland

[4] Department of Civil, Chemical, Env. and Materials Engineering, University of Bologna, Italy

[a] up201306134@fe.up.pt, [b] jacorreia@fe.up.pt

Keywords: Fatigue Damage, Hot-Spot Stresses, Notch Strain, Offshore Structures

Abstract. In offshore structures, the consecutive environmental and operational loading lead to an ever-changing stress state in the topside structure as well as in the substructure, which for offshore jacket-type platforms (called of fixed offshore structures) commonly used, result in fatigue damage accumulation. A wide variety of codes and recommended practices provide approaches in order to estimate the fatigue damage in design phase and remaining life in existing structures. In this research work, fatigue damage accumulation analyses applied to an offshore jacket-type platform using hot-spot stress and notch strain approaches are presented. These analyses are performed using wave information from the scatter diagram collected in North Sea. The wave loads used in this analysis were obtained using the Stokes 5th order wave theory and Morrison formula. The jacket-type offshore structure under consideration has a total height of 140.3 meters, a geometry at mud line of 60×80 meters and composed by tubular elements.

Introduction

Due to their location and function, offshore structures have undergone a significant improvement over the years converging in innovative solutions and materials to tackle the problems. With constant and/or variable cyclic loading from the environment, namely, environmental loads, this kind of structures are subject to fatigue damage accumulation resulting in the appearance of fatigue cracks leading to a reduction in their service life.

Several research studies to evaluate fatigue damage accumulation in offshore structures for oil & gas extraction and to support wind turbine towers have been proposed [1-4]. Michalopoulos and Zaaijer [1] have developed studies to assess the fatigue damage based on simplified approaches applied to offshore wind support structures accounting for variations in an offshore wind farm caused by wind and wave loading. Siriwardane et al. [2] proposed an accurate fatigue damage model for offshore welded joints subjected to variable amplitude loading. This new model is based on damage transfer concept using only the S-N curve given in the standard codes of practice. Kajolli and Siriwardane [3] have proposed a new approach for estimating fatigue life in offshore steel structures based on a sequential law as well as in the hot-spot stress approach accordingly DNVGL-RP-C203 code [4]. Conti et al. [5] proposed a fatigue assessment of tubular welded connections with the structural stress approach and considering the Dang Van criterion. A comparison with traditional hot-spot stress approach was made and showed that the proposed methodology is consistent with the existing approach and enables the consideration of the

Experimental Mechanics of Solids   Materials Research Forum LLC
Materials Research Proceedings 12 (2019) 146-154   https://doi.org/10.21741/9781644900215-21

beneficial effects of compression in legs, which was already observed in tubular joints tested under compression.

The fatigue design of offshore steel structures is normally made based on recommendations from the DNVGL-RP-C203 code [4]. The fatigue life may be calculated based on the S-N fatigue approach under the assumption of linear cumulative damage Palmgren-Miner method [6]:

$$D = \sum_{i=1}^{k} \frac{n_i}{N_i} = \frac{1}{\bar{a}} \sum_{i=1}^{k} n_i \cdot (\Delta\sigma_i)^m \leq \eta. \tag{1}$$

where, D is the accumulated fatigue damage, $\bar{a}$ is the intersection of the design S-N curve with the log N axis, m is the negative slope of the S-N curve, k is the number of stress blocks, n_i is the number of stress cycles in stress block i, N_i is the number of cycles to failure at constant stress range $\Delta\sigma_i$, and η is the usage factor (1/Design Fatigue Factor). The DFF parameter (Design Fatigue Factor) to be used in fatigue design and analysis is based on classification and accessibility to the structural component [7].

Alternatively, a simplified fatigue analysis based on a long-term stress range distribution may be presented as a two-parameter Weibull distribution, which is also presented in the DNVGL fatigue design code [4]:

$$D = \left(\frac{n_0}{a}\right) \cdot q^m \cdot \Gamma\left(1 + \frac{m}{h}\right) \leq \eta. \tag{2}$$

where, n_0 is the number of occurrences, q and h the Weibull distribution scale and shape parameters, respectively, and Γ the gamma function.

In this study, fatigue damage accumulation analyses based on traditional and simplified probabilistic fatigue methodologies using notch strain and hot-spot stress approaches, respectively are presented and applied to an offshore jacket-type platform. In this way, the wave characterization aiming at simulating and posteriorly achieve the structural response is required. Wave measurements in the North Sea were made to obtain the scatter diagram, which correlates wave periods, wave height, and the number of occurrences. The wave loads to be applied in elements of the structural model were obtained using the fifth order Stokes wave theory [8] and the Morison's formula [8].

The offshore jacket-type platform under consideration has a height of 140.3 meters (an elevation of 27 meters above sea level and a water depth of 113.3 meters) and geometries at mud line and top side interface of 60×80 meters and 24×80 meters, respectively. All members of the offshore structure were built using tubular elements of S420 structural steel [9,10].

Hot-spot and Notch Stresses Approaches
The fatigue design codes have suggested the use of nominal, hot-spot and notch strain approaches to find the most efficient S-N design curve of a considered structural detail. In Figure 1, it can be seen the nominal, hot-spot and notch stresses in a tubular welded joint [5]. Very often, the notch stress due to the local weld geometry is excluded from the stress calculation [4,6]. In this situation, hot-spot stresses are assumed to generate design S-N curves.

Experimental Mechanics of Solids   Materials Research Forum LLC
Materials Research Proceedings **12** (2019) 146-154   https://doi.org/10.21741/9781644900215-21

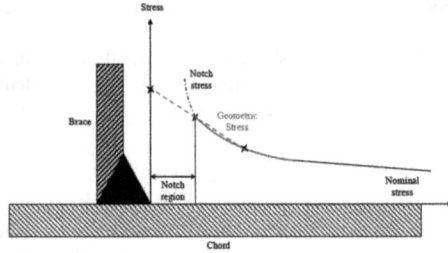

Fig. 1. Notch region and geometric hot-spot stress definition [5].

Hot-spot Stress approach
The fatigue damage assessment for tubular welded joints according to DNVGL-RP-C203 code is based on hot-spot stress approach instead of a nominal and notch stress approaches.

The design value of geometrical (hot-spot) stress, σ_{hs}, should be determined as follows [4,6]:

$$\sigma_{hs} = SCF \cdot \sigma_{nom}. \tag{3}$$

where, σ_{nom} is the nominal stress and SCF is the stress concentration factor calculated by Efthimiou' formulas for tubular welded joints based on geometrical parameters and loading modes (axial, in plane bending, out of plane bending) [4]. Once the stress concentration factors for axial, in-plane-bending and out-of-plane bending are obtained, the hot-spot stresses according DNVGL-RP-C203 code [4] are determined based on a superposition of the stress concentration factors at 8 different points for different loading modes at the weld toe around the tubular joint correlated with the nominal stresses. Alternatively, the hot-spot stresses can be evaluated by a finite element simulation and considering the stress distribution through a two-point linear regression at $t/2$ and $3/2t$ distances from the weld toe, where t is the thickness of the tubular element under consideration [11].

Notch Stress approach
Notch stress is commonly referred as local- or peak stress, which occurs at the weld toe or at critical point in a holed plate. This stress include the notch effect of a structural detail which occurs along the notch zone until the beginning of the weld toe.

The design value of local (notch) stress, σ_{loc}, should be determined as follows [4,6]:

$$\sigma_{loc} = SCF \cdot \sigma_{nom}. \tag{4}$$

where, σ_{nom} is the nominal stress and SCF is the stress concentration factor. The notch stress approach have a counterpart notch strain approach that is advantageous when some level of plasticity could appear. In case of fully elastic behavior both notch stress and strain approaches are equivalent.

Proposed Fatigue Methodologies Based on Hot-Spot and Notch Strain Approaches
In Fig. 2 and 3, the fatigue methodologies based on hot-spot and notch strain approaches are presented. The hot-spot stresses, around of weld toe of the tubular joints, are obtained using the Efthymiou equations with aims to determine the stress concentration factors (SCF), which are correlated with the nominal stresses for different loading conditions according to the DNVGL-

Experimental Mechanics of Solids
Materials Research Proceedings **12** (2019) 146-154

Materials Research Forum LLC
https://doi.org/10.21741/9781644900215-21

RP-C203 code. Simplified fatigue method proposed in the same standard is used in the proposed fatigue procedure presented in Figure 2. The second proposed fatigue methodology based on notch (local) strain approach uses the Neuber rule and Ramberg-Osgood description as well as the Coffin-Manson relation with objective to evaluate the strain amplitude and number of cycles [12,13].

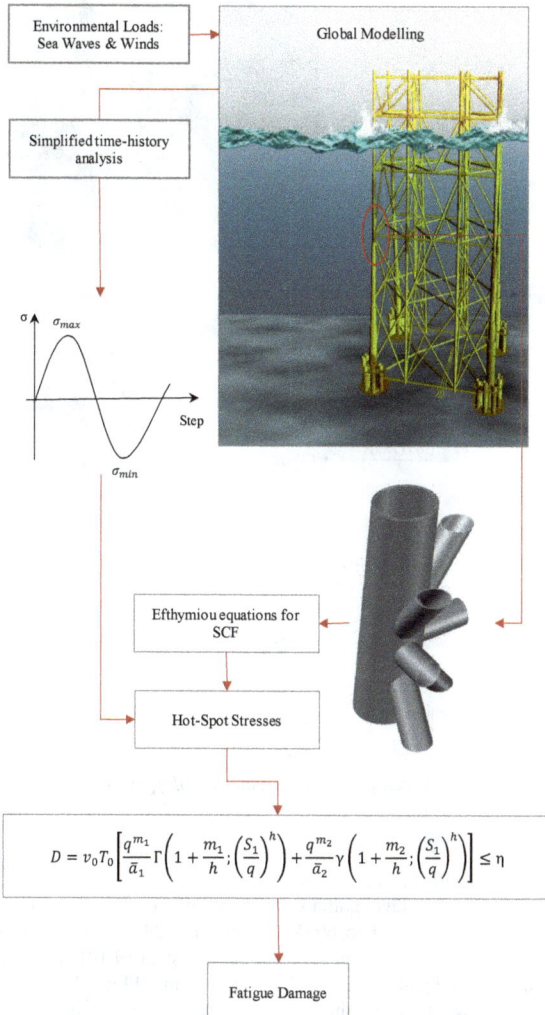

$$D = v_0 T_0 \left[\frac{q^{m_1}}{\bar{a}_1} \Gamma \left(1 + \frac{m_1}{h}; \left(\frac{S_1}{q} \right)^h \right) + \frac{q^{m_2}}{\bar{a}_2} \gamma \left(1 + \frac{m_2}{h}; \left(\frac{S_1}{q} \right)^h \right) \right] \leq \eta$$

Fig. 2. Hot-spot stresses methodology workflow.

Experimental Mechanics of Solids   Materials Research Forum LLC
Materials Research Proceedings **12** (2019) 146-154   https://doi.org/10.21741/9781644900215-21

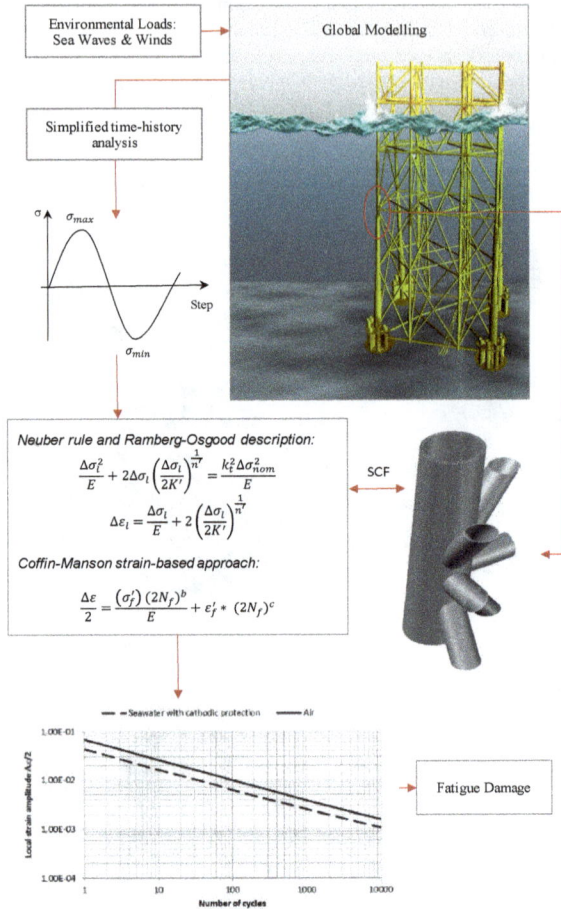

Fig. 3. Notch (local) strain methodology workflow.

Application and Results

The offshore jacket-type platform under consideration has a total height of 140.3 meters – an elevation of 27 meters above sea level and a water depth of 113.3 meters. The geometries at mud line and top side interface are equal to 60×80 meters and 24×80 meters, respectively, and all members are built in tubular elements of S420 structural steel [9,10]. The offshore structure is composed by horizontal bracing at elevation -108.9 m, -73 m, -44 m, -15 m, +8 m, and +24 m. In Fig. 4 and 5, an overview of the offshore jacket-type platform under consideration and the critical joint can be seen [9,10].

The fatigue analysis of the offshore jacket-type platform was made considering the wave loads and ignoring the wind loads. The fatigue damage due to wind loading when compared to

Experimental Mechanics of Solids   Materials Research Forum LLC
Materials Research Proceedings **12** (2019) 146-154   https://doi.org/10.21741/9781644900215-21

wave loading is so small that it is not considered in the analysis. Morrison forces according to 5th order Stokes wave theory for several water depths and for a wave with 14.5 m height can be seen in Fig. 6. Permanent loads will not contribute to fatigue damage and were also excluded from the analysis. The stress ranges from each wave force based on wave scatter diagram were obtained using the SESAM software. In this analysis 2304 load cases resulting from the 12 wave directions, 8 wave heights with corresponding periods, and 24 steps in the wave were considered. The dynamic response of the offshore structure was made using the SESAM software (see Fig. 7 to 9).

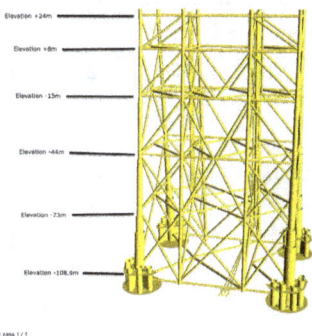

Fig. 5. Critical joint at elevation -44 m.

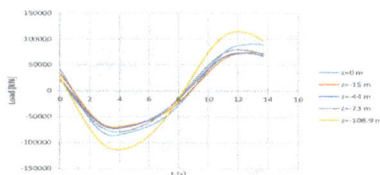

Fig. 4. Overview of the offshore jacket-type platform.

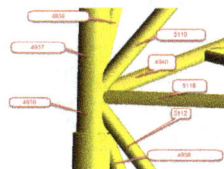

Fig. 6. Morrison force (14.5 m wave height).

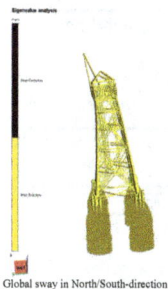

Fig. 7. Vibration mode 1: Period 3.18 seconds.

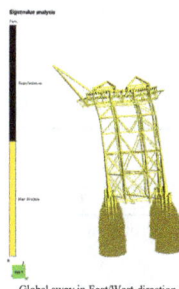

Fig. 8. Vibration mode 2: Period 2.99 seconds.

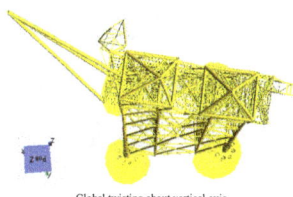

Fig. 9. Vibration mode 3: Period 1.90 seconds.

The fatigue methodologies presented in Figures 2 and 3 were used and the results are presented in Fig. 10. The fatigue methodology based on hot-spot stress approach (Fig. 2) was applied considering the stress concentration factors and geometrical dimensions of the critical joint at elevation -44 m determined using the DNVGL-RP-C203 recommendations (see Table 1). The design fatigue S-N curve for tubular joints in a seawater environment with cathodic protection according to DNVGL-RP-C203 code was used. In the other hand, for the fatigue methodology based on notch (local) strain approach, the cyclic properties of S420 steel based on

151

Experimental Mechanics of Solids
Materials Research Proceedings **12** (2019) 146-154

Materials Research Forum LLC
https://doi.org/10.21741/9781644900215-21

DNVGL-RP-C208 standard [14] as well as Coffin-Manson strain-life properties from ref. [15] (the strain-life parameters used in this analysis correspond to S355 structural steel) were used (see Table 2). In both methodologies, the Palmgren-Miner linear damage rule was used to determine the fatigue damage using the usage factor of 0.1 and service life of 50 years. In Figure 10, the summary of fatigue damage obtained for both analyses considering extreme waves from the scatter diagram called W73 (smallest wave height and most number of occurrences) and W80 (biggest wave height and less number of occurrences).

Table 1. Stress concentration factors and geometrical dimensions for members of the critical joint.

	Location	SCF_{BAL}	SCF_{IPB}	SCF_{OPB}	SCF_{UOPB}	Diameter [m]	Thickness [m]
	5110	1.6009775	1.2990365	1.8363963	2.5436804	1.2	0.04
Chord	5116	2.9490646	1.3706252	2.5918459	2.9185857	1	0.03
	5112	1.3688033	1	1.0828665	2.1077958	1.1	0.025
	5110	1.8233236	1.6235408	2.6893376	-	1.2	0.04
Brace	5116	3.4089698	2.1717246	3.7857918	-	1	0.03
	5110	1.7857226	1.5148	2.9438576	-	1.1	0.025
	4939	1.8318157	1	1.1413502	2.9810327	1.2	0.035
Chord	4940	5.4295072	2.3804706	5.6410504	5.762028	1.32	0.055
	4938	1.6207604	1	1.2818332	3.2921516	1.1	0.025
	4939	1.7222647	1.4198678	3.3873849	-	1.2	0.035
Brace	4940	4.5199659	2.4921638	4.9807336	-	1.32	0.055
	4938	2.0016058	1.6093898	4.5979906	-	1.1	0.025
Chord	4936	-	-	-	-	2.3	0.095
	4937	-	-	-	-	2.3	0.095

BAL – Balanced axial load; IPB – In-plane bending; OPB – Out-of-plane bending; UOPB – Unbalanced out-of-plane bending.

Table 2. Monotonic, cyclic and Coffin-Manson parameters of the S420 structural steel.

E GPa	f_u MPa	f_y MPa	K' MPa	n' -	σ'_f * MPa	b * -	ε'_f * -	c * -
210	561.2	426.3	690	0.1	952.2	-0.089	0.7371	-0.664

* Strain-life parameters of the S355 structural steel used in this study.

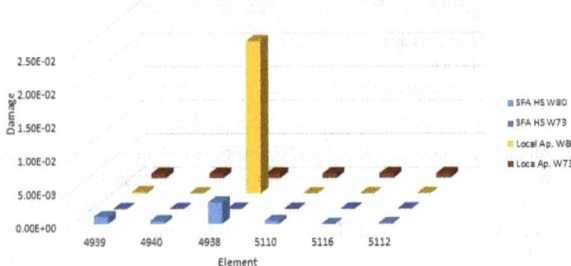

Fig. 10. Summary of fatigue damage obtained for both analysis.

Experimental Mechanics of Solids   Materials Research Forum LLC
Materials Research Proceedings **12** (2019) 146-154   https://doi.org/10.21741/9781644900215-21

Conclusions

All fatigue damage obtained in both analyses are inferior to the limit imposed to the usage factor of 0.1. The fatigue damage estimated using the methodology based on the notch (local) strain approach according to the Neuber rule and Ramberg-Osgood description shows a significantly higher damage value when compared with the methodology based on hot-spot stress approach. Experimental strain-life tests for the low- and high-cycle fatigue regimes are recommended.

Acknowledgments

Authors gratefully acknowledge the contribution of Force Technology for the distribution of the data used for this research. This work was also financially supported by: UID/ECI/04708/2019-CONSTRUCT - Instituto de I&D em Estruturas e Construções funded by national funds through the FCT/MCTES (PIDDAC).

References

[1] V. Michalopoulos and M. Zaaijer, "Simplified fatigue assessment of offshore wind support structures accounting for variations in a farm", European Wind Energy Association Annual Conference and Exhibition 2015, EWEA 2015 - Scientific Proceedings 2015.

[2] A. Aeran, S. Siriwardane, O. Mikkelsen, and I. Langen, "An accurate fatigue damage model for welded joints subjected to variable amplitude loading", in IOP Conference Series: Materials Science and Engineering, 2017, vol. 276, no. 1, p. 012038: IOP Publishing. https://doi.org/10.1088/1757-899x/276/1/012038

[3] R. Kajolli, "A new approach for estimating fatigue life in offshore steel structures", MSc. Thesis, 130 pages, University of Stavanger, Norway, 2013.

[4] DNV GL Group. DNVGL-RP-C203: Fatigue design of offshore steel structures, 2016.

[5] F. Conti, L. Verney, and A. Bignonnet, "Fatigue assessment of tubular welded connections with the structural stress approach", Fatigue Design 2009, 25-26 November 2009, Senlis, France, 8 pages.

[6] CEN-TC 250. EN 1993-1-9: Eurocode 3, Design of steel structures – Part 1-9: Fatigue. European Committee for Standardization, Brussels; 2003.

[7] DNV GL Group. DNVGL-OS-C201: Structural design of offshore units - WSD method, 2015.

[8] DNV GL Group. DNV-RP-C205: Environmental Conditions and Environmental Loads, 2014.

[9] A. Mourão. "Fatigue analysis of a jacket-type offshore platform based on local approaches", MSc Thesis, 193 pages, Civil Engineering, Faculty of Engineering, University of Porto, Porto, Portugal, 2018.

[10] António Mourão, J.A.F.O. Correia, J.C. Rebelo, M.Correia, N. Fantuzzi and R. Calçada, "Fatigue wave loads estimation using Morison formula for an offshore jacket-type platform", 19th International Colloquium on Mechanical Fatigue of Metals, Porto, Portugal, 2018.

[11] Recommendations for Fatigue Design of Welded Joints and Components. International Institute of Welding, doc. XIII-2151r4-07/XV-1254r4-07; Paris, France, October 2008.

[12] A. Fernández-Canteli, E. Castillo, H. Pinto, and M. López-Aenlle, "Estimating the S–N field from strain–lifetime curves," Strain, vol. 47, pp. e93-e97, 2011. https://doi.org/10.1111/j.1475-1305.2008.00548.x

Experimental Mechanics of Solids
Materials Research Proceedings 12 (2019) 146-154

Materials Research Forum LLC
https://doi.org/10.21741/9781644900215-21

[13] A. M. De Jesus, H. Pinto, A. Fernández-Canteli, E. Castillo, and J. A. Correia, "Fatigue assessment of a riveted shear splice based on a probabilistic model," International Journal of Fatigue, vol. 32, no. 2, pp. 453-462, 2010. https://doi.org/10.1016/j.ijfatigue.2009.09.004

[14] DNV GL Group. DNVGL-RP-C208: Determination of Structural Capacity by Non-linear finite element analysis Methods, 2016.

[15] F. Öztürk, J. Correia, C. Rebelo, A. De Jesus, and L. S. da Silva, "Fatigue assessment of steel half-pipes bolted connections using local approaches," Procedia Structural Integrity, vol. 1, pp. 118-125, 2016. https://doi.org/10.1016/j.prostr.2016.02.017

Experimental Mechanics of Solids
Materials Research Proceedings 12 (2019) 155-158

Materials Research Forum LLC
https://doi.org/10.21741/9781644900215-22

The Influence of Bonding System Type on Strength Properties of Reconstructed Teeth with 1st Class Cavities According to Black's Classification

Grzegorz Milewski[1,a] *, Agata Borowiec[1,b]

[1]Cracow University of Technology, Institute of Applied Mechanics, Warszawska 24, 31-155 Cracow, Poland

[a]milewski@mech.pk.edu.pl, [b]borowiec.agata@gmail.com

Keywords: Dental Bonding System, First Class Tooth Cavity Reconstruction, Strength Properties

Abstract. The paper presents the results of strength tests of lateral teeth with 1st class cavities reconstructed with various bonding systems, i.e. self- and selective bonding systems as well as for filling with additional etching with orthophosphoric acid. In all cases a nano-hybrid, light-cured composite material was used in a layered (sandwich) cavity filling processing. The strength tests were done with the use of Instron 4465 strength machine for loadings simulating the features of proper lateral teeth occlusion. The analyses of the results were done with respect to the following quantities: ultimate force, total displacement and work to fracture. The results were compared with reference 'healthy' lateral teeth group. Analysis of the results proves that the type of bonding system has a significant effect on the values of individual strength parameters of the tested tooth samples. From the mechanical point of view the best results were obtained for self- and selective bonding system when comparing with additional etching procedure. However, regardless on the type of bonding system, the mechanical properties of teeth subjected to dental treatment are significantly lower than those for the 'healthy' teeth.

Introduction

Adhesion is the basic physical phenomenon that determines the success of composite crown fillings in contemporary conservative dentistry [1]. Phenomenon of adhesion consisting in the combination of surface-contact layers of two bodies in a solid or liquid phase combines with the mechanical, polarizing, diffusive and adsorptive theories of the mutual interaction of two bodies. Close contact of the contacting bodies is necessary for the formation of the adhesion phenomenon - the distance between the bodies should be smaller than 1 - 2 Å (10 – 20 nm) – Fig. 1a. In practice, achieving such a low distance between solid bodies is difficult to achieve. Therefore, in order to obtain adhesion forces, a liquid photo polymerized material is used to wet the substrate. This material, called a bonding system, acts as an intermediate adhesive layer and provides better connection with dental fillings. The bonding systems have been developed over years. The differences between all generations of bonding systems are significant. Development of materials and techniques have changed the bonding systems themselves as well as their adhesive properties and strength [2]. In general, adhesion is determined by the value of the unitary adhesion forces or the work necessary to separate the adherent bodies, however various physical quantities are used to quantify that phenomenon between hard tissues of tooth (i.e. dentine and enamel) and dental composite filling material. The aim of the work was to determine the strength properties of reconstructed lateral tooth crowns with the use of various bonding systems.

Materials and Methods

Black's scale is the most often used classification in contemporary conservative dentistry. It is based on the tooth type and the cavity location or tooth surfaces involved. Black's scale consists of (I – VI) classes [1]. The first class cavities are located in pits or fissures of crowns, mainly in the occlusal surfaces of molars and premolars as well as in the lingual surfaces of upper incisors, and occasionally in the lingual surfaces of upper molars – Fig. 1b.

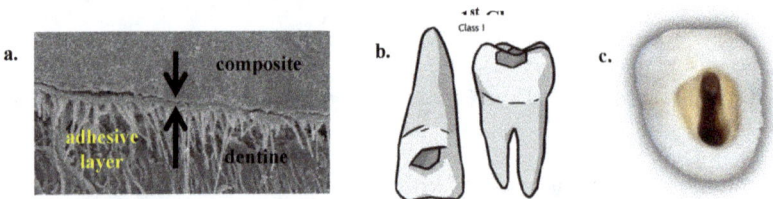

Fig. 1. SEM image of adhesive layer between dentine and composite filling material (a); scheme of 1ˢᵗ class cavity according to Black's classification (b) and tooth crown cavity preparation after endodontic treatment ready for final reconstruction (c)

The tests were done for 15 samples - lateral teeth after extraction for orthodontic reasons. All teeth were prepared for the 1ˢᵗ class cavities according to Black's classification. The teeth were randomly divided into three reconstructive groups (5 samples in each group): I - teeth reconstructed with self-bonding system that does not require additional etching procedure, II – reconstruction with additional etching with orthophosphoric acid, III – teeth reconstructed with the use of a selective-etching bonding system – Fig. 2. All teeth-samples were prepared under a supervision of a dentist and were done with standard dental treatment procedures. A nano-hybrid, light-cured composite material was used in the reconstruction of all teeth crowns, which were worked out with layered (sandwich) cavity filling processing [1, 3].

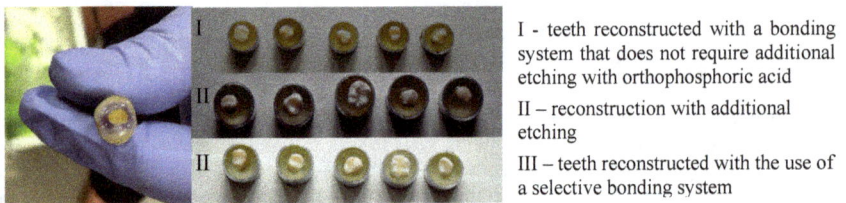

I - teeth reconstructed with a bonding system that does not require additional etching with orthophosphoric acid

II – reconstruction with additional etching

III – teeth reconstructed with the use of a selective bonding system

Fig. 2. Set of tooth samples with crowns reconstructed with the use of various bonding systems

The tests were done with the use of Instron 4465 strength machine for loadings simulating the features of proper teeth occlusion. As there is no standardized method of strength tests of tooth crowns a compression test with a steel ball that is used instead of an opposing tooth was applied. The method was proposed and developed in [4, 5, 6]. The following quantities were registered and analyzed: ultimate force F_c , work to fracture W_c , total displacement u_c . The results of the

Experimental Mechanics of Solids
Materials Research Proceedings **12** (2019) 155-158

Materials Research Forum LLC
https://doi.org/10.21741/9781644900215-22

strength tests in all groups were compared with the respective results for reference 'healthy' lateral teeth (group 0) strength tests which were done in the previous author's work [4].

Results and Discussion

The results of strength tests carried out for application of three different bonding systems for 1st class cavities in lateral teeth treatment are presented in Table 1. The data are given for mean values and standard deviations (SD) for each respective teeth groups.

Table 1. Set of registered and analyzed values of ultimate force F_c, total displacement u_c and work to fracture W_c for lateral teeth groups reconstructed with various bonding systems

Teeth group	Ultimate force [kN]		Displacement [mm]		Work to fracture [J]	
	mean	SD	mean	SD	mean	SD
0	1.850	0.1996	2.162	0.1032	1.530	0.0873
I	1.005	0.2849	1.433	0.7116	0.599	0.3969
II	0.563	0.1783	1.714	0.7940	0.291	0.0854
III	0.947	0.1486	1.614	0.7399	0.574	0.1910

The analysis consisted of comparison of all specimens data in each group and also of comparison of mean values of each parameter in the selected groups. All used types of bonding systems were compared with each other. The analyses were done with regards to the respective values obtained for the 'healthy' teeth. The obtained results allow to conclude that all cavities reconstructions, regardless the bonding system type, decrease the mechanical properties of teeth. The smallest differences between the destructive forces (F_c) occur for self-bonding system (group I). The difference between these values is around 45%. Similar values were obtained for the third research group, here the difference is about 48%, while the difference between the respective values of ultimate forces for the system requiring additional etching (group II) is almost 70%. For the work to fracture (W_c) the relevant ratios are even higher and reaches 61%, 64% and 81% respectively, what stands for the essential reduction of the ability to energy dissipation for lateral teeth crowns subjected to dental treatment. This regularity is illustrated by a histogram presented in Fig. 3.

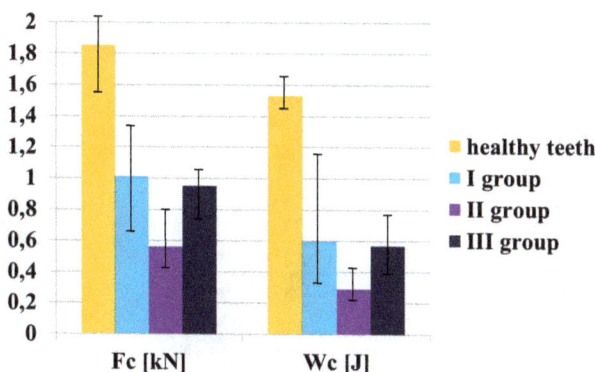

Fig. 3. Comparison of the mean values of ultimate force F_c and work to fracture W_c for lateral teeth crowns reconstructed with the use of various bonding system

Summary
Analysis of the results proves that the type of bonding system has a significant effect on the values of individual strength parameters of the tested tooth samples. The best results were obtained for self- and selective-etching bonding system. However, regardless the type of bonding system, the mechanical properties of teeth subjected to dental treatment are significantly lower than those for the 'healthy' teeth. Samples damage analysis shows that in general there was no delamination of the composite material what proves the good strength properties of all bonding systems used in the research. The study shows that there is a need for further development of bonding systems and dental fillings in order to increase their mechanical properties.

References

[1] R. Ireland (Ed.), A Dictionary of Dentistry, Oxford University Press, Oxford, New York, 2010

[2] R. L. Sakaguchi, J. M. Powers (Eds.), Craig's Restorative Dental Materials, 13th Edition, Elsevier/ Mosby, Philadelphia, 2012

[3] G. Milewski, T. Majewski, Influence of the method of polymerization of composite material for teeth crown fillings on its strength properties, Solid State Phenomena (2016), 240, pp. 168-173. https://doi.org/10.4028/www.scientific.net/ssp.240.168

[4] G. Milewski, Numerical and experimental analysis of effort of human tooth hard tissues in terms of proper occlusal loadings, Acta of Bioengineering and Biomechanics (2006), 7 (1), pp. 47-58

[5] R. Sorrentino, Z. Salameh, F. Zarone, F.R. Tay, M. Ferrari, Effect of post-retained composite restoration of MOD preparations on the fracture resistance of endodontically treated teeth, Journal of Adhesive Dentistry (2007), 9 (1), pp. 49-56

[6] G. Milewski, A. Hille, Experimental strength analysis of orthodontic extrusion of human anterior teeth, Acta of Bioengineering and Biomechanics (2012), 14(1), pp. 15-21

Experimental Mechanics of Solids
Materials Research Proceedings **12** (2019) 159-165

Materials Research Forum LLC
https://doi.org/10.21741/9781644900215-23

Multi-phase Magnetostrictive Actuator Dedicated to Close-Loop Vibration Control System

Jerzy Kaleta[1,a], Karol Wachtarczyk[1,b] and Przemysław Wiewiórski[1,c *]

[1]Wroclaw University of Technology, Department of Mechanics, Materials Science and Engineering, Smoluchowskiego str 25, 50-370 Wroclaw, Poland

[a]jerzy.kaleta@pwr.edu.pl, [b]karol.wachtarczyk@pwr.edu.pl,
[c]przemyslaw.wiewiorski@pwr.edu.pl

Keywords: Magnetostrictive Actuator, Multi-Phase Vibrations, Moving Objects Using Vibrations, Digital Signal Processing, Code Of Structure Stiffness (CSS)

Abstract. The article discusses a multi-phase magnetostrictive actuator which is the executive part of a multi-phase mechanical vibration controller (MVC). The whole system contains several stages of magnetostrictive actuators and a dedicated vibration sensor. Each individual actuator is dedicated to operation with the same frequency and with an individually programmed phase shift. The controller enables the actuator to excite mechanical vibration in various media (including liquids) to ultrasound frequencies. This allowed to continuously maintain a selected construction in mechanical resonance whose frequency was determined in real time. The work is the continuation of research on the development of the Energy Harvesting (EH) methods by using magnetostrictive actuator - mechanical cross phenomena.

Introduction

The main objective of this work was to construct the multi-phase magnetostrictive actuator forming the executive part of a multi-phase mechanical vibration controller. It was assumed that the proposed solution would enable positioning of objects freely placed on the base (beam) subjected to mechanical vibration. Simultaneously it was assumed that in constructions there is such a form of vibration, up to 30 kHz, which allows to supply energy in a controlled way and perform remote mechanical work (e.g. displacement in a set direction, mechanism revolution, unscrewing an element, pressing a switch). For this purpose the idea of generating equal frequency phase-shifted vibration. It involves the excitation of actuators distributed one after another in such a way that subsequently sent signals are shifted with regards to one another by a set angle, which is changed in time in accordance with the adopted algorithm.

Due to the feedback from an integrated vibration sensor, the multi-phase method of shaping mechanical vibration allowed to define and use the so called Code of Structure Stiffness (CSS), which is described below.

However, to achieve the goals the following tasks had to be performed:

- making a prototype of the head of a multi-phase actuator with an integrated vibration level sensor,
- making a dedicated electronic system, vibration controller software and the determination of actuator performance,
- defining the so called Code of Structure Stiffness – CSS,
- developing an original method of objects and mechanisms positioning on the basis of the free displacement of an element along a straight line beam undergoing vibration.

In the summary further development directions were suggested.

Experimental Mechanics of Solids   Materials Research Forum LLC
Materials Research Proceedings **12** (2019) 159-165   https://doi.org/10.21741/9781644900215-23

Magnetostrictive exciters of mechanical vibration

For the purpose of generation of mechanical vibrations is needed to understand of various types of vibrators so called vibration exciters. Their selection depends on vibration level (PSD – Power Spectral Density), frequency and amplitude-phase level of distortion. If it is required that actuator doesn't contain moving parts, the selection is limited to only two types of transducers: Piezoelectric Lead-Zirconate-Titanate (PZT) and a magnetostrictive actuator (most often using Terfenol-D). It should be noted, however, that piezoelectric generators operate only in a precisely determined, narrow resonant frequency range related to the construction of the piezoelectric transducer. A different frequency range results in actuator overheating or PZT rings destruction. On the other hand magnetostrictive actuators operate in a wider frequency range, including also the resonance of the vibrator's construction itself [1] and are characterised by a large generated force (reaching hundreds of kN) and high maximum work frequency (up to even 100kHz) [2]. Their disadvantage is non-linear characteristic, low vibration amplitude (compare to electrodynamic tranducers) and maximum work temperature, restricted by the acceptable parameters of the induction coil. The price of magnetostrictive actuators is much higher than in the case of typical piezoelectric transducers, hence their use is limited to a small number of applications. However, the predominance of advantages was the reason why magnetostrictive actuators were used in this solution.

The main element in magnetostrictive exciters is the so called magnetostrictive actuator core which is composed of one or more rods, depending on its length, made of materials characterised by giant magnetostrictiction (GMM - Giant Magnetostrictive Materials, e.g. Terfenol-D, cobalt nano-ferrite) and a system adjusting the magnetization level on the basis of neodymium magnets. The number of core components (GMM rods, neodymium magnets, pressure mandrels) is of key significance for its construction, the fewer of them, the higher work frequencies are achieved. One of the solutions used in the core is a system made of alternately located neodymium magnets and Terfenol-D rods [3].

The basic characteristics of materials with giant magnetostriction significantly depend on the parameters of external fields. Due to this the actuator constructions are equipped with permanent magnets generating a bias magnetic field and a spring system causing prestress. The appropriate prestress level allows to increase magnetostrictive amplitude [4,5]. The bias magnetic field in turn increases the linearity of characteristics and sensitivity [6].

The authors have considerable experience in constructing actuators and magnetostrictive actuator harvesters [3], and their earlier research on the use of Terfenol-D have found a number of applications in exciters and vibration transducers [2,7,8].

Construction of vibration level based on a multi-phase magnetostrictive actuator and the Heron system

The controlled generation of mechanical vibration in a wide range of frequencies was conducted using a multi-phase magnetostrictive actuator with an integrated sensory part. The whole system was controlled with Heron Advanced Multiphase Software which provides control signals to the actuator (frequency and vibration phases) required to maintain the construction in resonance. The vibration controller uses an electronic system containing DSP (Digital Signal Processor) and measurement modules, i.e. input-output modules, DDS (**Direct Digital Synthesis**) generators and an ICP (Integrated Electronics Piezo-Electric) sensor from PCB Piezotronics. Signal conditioning and increasing level of vibration power is ensured thanks to a dedicated CDM-1P device.

The generation of the so called multi-phase vibration requires a head containing numerous active magnetostrictive parts. In the case of the presented research a head made of four actuators was constructed due to the number of Terfenol-D rods (8 pieces) and the possibility to use

Experimental Mechanics of Solids   Materials Research Forum LLC
Materials Research Proceedings **12** (2019) 159-165   https://doi.org/10.21741/9781644900215-23

analogue control as in a typical stepper motor. Moreover, a PCB vibration sensor was used centrally in the system to measure the level of vibration in the excited object. The proposed distribution of actuators and the sensory part is presented in Fig. 1.

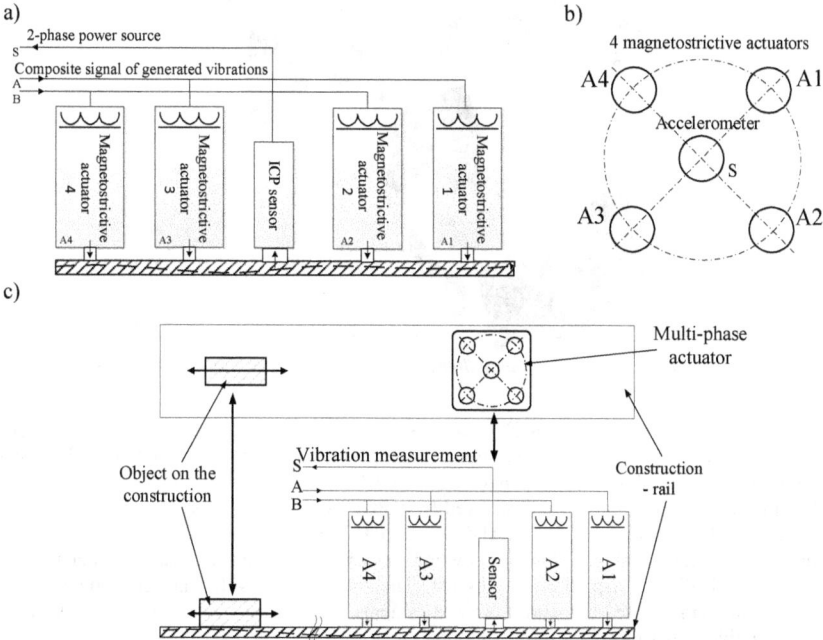

Fig. 1. Idea of the system of four magnetostrictive actuators: a) scheme of actuators operating simultaneously with vibration measurement path; b) symmetrically distributed actuators and central vibration sensor; c) scheme of linear positioning system using vibration

The design and prototype of the head was made in accordance with Fig. 1. Actuators were situated symmetrically (Fig. 1B). Figure 1C shows a scheme of the positioning system using generated vibration with feedback in the case of a straight line bus which is described below.

Figure 2 presents the whole vibration controller system with the designed multi-phase head. Every actuator can be operated separately. It allows to generate even complex mechanical vibrations modes. Apart from the head, the vibration controller system was composed of advanced electronic subsystems based on a Heron card with a floating-point DSP, Texas Instruments C6000 type with extending modules. The system contained also dedicated software using API, Hunt Engineering.

Experimental Mechanics of Solids   Materials Research Forum LLC
Materials Research Proceedings **12** (2019) 159-165   https://doi.org/10.21741/9781644900215-23

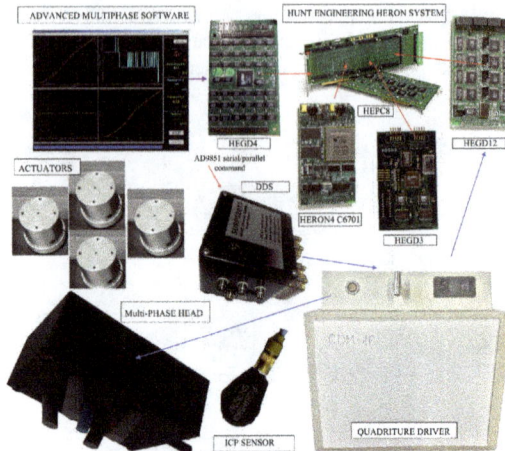

Fig. 2. View of complete vibration controller system based on Hunt Engineering DSP Heron system [9]

System Hunt Engineering HERON was responsible for the signal processing – from acquisition and conditioning the sensor signal to the generation of phase shifted control signals using both DAC (Digital-to-Analog Converter) and DDS stages.

Vibration as electric signals were received by a PCB sensor after appropriate suppling they were collected by a module of 16-bit ADCs - HEGD12. They were used as the basis for the determination of the setpoints of the digital vibration controller which released subsequent DDS values in the feedback loop through HEGD4. The role of CDM-1P was only conditioning (amplification and noise reduction) sensor signals and power amplification for four magnetostrictive actuators.

The Heron Advanced Multiphase Software was to maintain the construction in the resonant state despite the changing resonant frequency. For this purpose, it generated control signals for actuators with the required resonant vibration frequency. The determination of resonant vibration frequency was possible on the basis of the analysis of signal amplitude with a set frequency moment of time. If at a given excitation, the vibration amplitude was decreased, it meant that the system was "leaving" the resonant state. Then the algorithm checked whether after the excitation of the system with frequency higher or lower than the current vibration amplitude it grew. If so, this frequency was considered a new resonant frequency. The amplitude control cycle was repeated.

Remote object positioning using vibration
A test task used for this system was positioning an object located on the construction using vibration. The task was performed by generating vibration in the range of the mechanical resonance of the construction, and next placing a non-magnetic object, about 30 g, on a vibrating 6-meter long steel beam. Under the influence of mechanical vibration the object was set in motion resulting in subsequent hitting the beam structure. By changing the phase shift between generated signals, it was possible to position the object. The system scheme is presented in Fig. 3A.

The acceleration sensor fixed in the multi-phase actuator head recorded acoustic events occurring at the moment when the object separated from the beam. Its small mass only

Experimental Mechanics of Solids
Materials Research Proceedings **12** (2019) 159-165

Materials Research Forum LLC
https://doi.org/10.21741/9781644900215-23

insignificantly changes structure frequency and the evoked acoustic event had significantly higher frequency than the resonant frequency of the beam. Vibration changes were also recorded as a sinusoidal signal with a frequency of 667 Hz. When an aluminium object was placed on the beam, additional acoustic events resulting from the loss of contact between the object and the beam occurred.

Code of Structure Stiffness – CSS

The acoustic events resulting from hitting the beam with a small mass constitute a characteristic signal reply to of a given structure. They have different amount of energy accumulated in time, characteristic for the beam structure and the adhering element, i.e. small aluminium mass. This response may be a control signal for vibration controller. On the other hand, the possibility to control the vibration level allows to detect changes in the construction.

In other words the number and character of recorded acoustic events is related to the energy supplied to the tested structure in the form of mechanical vibration and the instantaneous state of this structure. Hence a sequence of acoustic events, as a modulated binary run (in accordance with F2F - frequency/double frequency or MFM - modified frequency modulation) will depend on the changing parameters of the medium in which vibration is propagated, including propagation obstacles. The following factors are of key significance here: mechanical structure specific stiffness, number of elements in a structure, induction frequency characteristic for any of the construction elements, material defects, e.g. pores or cracks. This dependence between the energy supplied in the form of vibration to the object and the obtained sequence of acoustic events is called the Code of Structure Stiffness (CSS). In other words, CSS is a characteristic change of frequency response of a complex multi-phase system resulting from the supplied energy. Using the CSS idea and the correct interpretation of the binary signal allows to operate the vibration controller in a wide frequency range that it is possible to relocate an object in a programmable way.

An example of how the CSS method is used is presented in Fig. 3. A series of acoustic events in an ICP sensor signal was distinguished as a result of positioning a shifted object. The essence of the CSS method is the determination of a binary series on the basis of the duration of acoustic events and appropriate the corresponding energy released to the system by the moving object. The amount of this energy is determined by the duration time of the high level form the initiation of a given event. In this way it is possible to obtain a unique code related to the condition of a mechanical construction, which will be the subject of further work.

The number of events was characteristic for the current state of the tested object. With the system of a vibration controller in a wide frequency band, it is possible to identify the dynamic characteristics of a mechanical construction by the classification of its "code" as a result of programmable pure-sinusoidal excitation.

The issue requires more extensive research, however, according to the authors a unique technique for positioning and assessment of objects has been developed. The measurement technique resulting from the method potentially may be used in SHM (Structural Health Monitoring) and also as a competitive method in vibroacoustics.

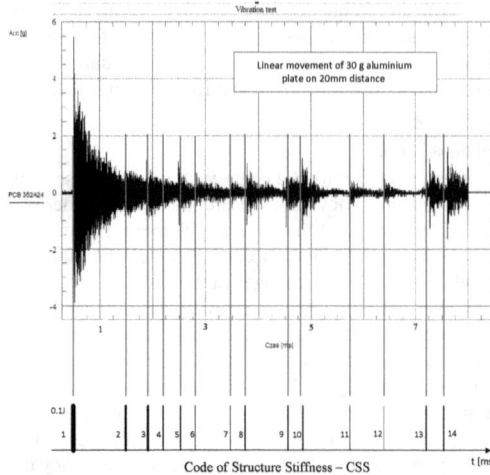

Fig. 3. Decoded signals from ICP sensor as energy stamps of acoustic events in vibration positioning system [9]

Further work directions – changes of the position of medical tools using programmable mechanical vibration.

According to the authors, the presented method, apart from its use in construction mechanics, can be also used in medicine to set the appropriate position of medical transducers placed in patients' bodies without the necessity for reoperation. Currently there are a number of solutions in which the calibration of medical transducers inside patients' bodies use a few physical phenomena, including: exposure to light of particular wavelength, electric and magnetic field. An example of a mechanism set using an external magnetic field is a valve regulating the level of intracranial pressure in intracranial hypertension (IH) symptoms. After remote programming the valve opening pressure (i.e. the angular position of the mechanism) the correctness of the setpoint is verified with an X-ray image [10].

The proposed development solution in medicine is the use of an acoustic wave with set energy density and a determined frequency band to perform similar "setting" tasks. Thanks to this it will be possible to avoid the influence of a strong magnetic field on mechanism components and also large heat losses which can lead to overheating patient's tissues.

Conclusions

The main goal of the work has been fully achieved, namely:

- A new generation head prototype was constructed on the basis of four magnetostrictive actuators in a symmetrical system integrated with an accelerometer.
- It was shown that the excitation of resonant vibration using the designed four-step head can be used to set in motion elements located on the surface of the analysed construction by setting the appropriate form of vibration becomes possible by their positioning.
- The so called Code of Structure Stiffness – CSS was defined and used. It was possible thanks to the use of a programmable mechanical vibration controller in a strictly defined frequency range. CSS can be used to classify the behaviour of the components of tested objects.

Experimental Mechanics of Solids   Materials Research Forum LLC
Materials Research Proceedings **12** (2019) 159-165   https://doi.org/10.21741/9781644900215-23

In future works the presented results will be investigated with special focus on mechanics application (active vibration damping, energy harvesting with transmission of power and data). One of the main research scopes will be the usage of controlled vibrations made by the designed actuator in NDE diagnostics. The presented method is the expansion of the methods known from Energy Harvesting due to the possibility to transform various forms of energy supplied to/collected from the system for the purpose of performing mechanical work which is object positioning.

The multi-phase actuator system will also be investigated as a potential method to be used in medicine for the positioning of medical devices placed in the patient's body without a need for reoperation. There are already plenty of methods for external calibration of medical transducers in the patient's body, but the proposed method can be used in this situation as well.

Moreover in further steps it will be necessary to pay attention to the costs of material used to make the core of the magnetostrictive actuator. Cores made of Terfenol-D are very expensive and it is legitimate to use another less costly composite replacement which will not decrease actuator parameters.

References

[1] F. Claeyssen, N. Lhermet, T. Maillard, Magnetostrictive Actuators Compared To Piezoeletric Acutators, 2002.

[2] ETREMA Products, Inc., 2012. Website: http://etrema-usa.com/.

[3] J. Kaleta, R. Mech and P. Wiewiórski, Development of Resonators with Reversible Magnetostrictive Effect for Applications as Actuators and Energy Harvesters, IntechOpen. https://doi.org/10.5772/intechopen.78572

[4] J. Bomba and J. Kaleta, An initial investigation into the change in magnetomechanical properties of Terfenol-D rod due to prestress and temperature, Issue of Material Testers, 2004.

[5] J. Bomba and J. Kaleta, The influence of prestress on magnetomechanical damping in giant magnetostrictive materials, w 20th DANUBIA-ADRIA Symposium on Experimental Methods in Solid Mechanics, Gyor, 2003.

[6] Y. Liang and X. Zheng, Experimental researches on magneto-thermo-mechanical characterization of Terfenol-D, Acta Mechanica Solida Sinica, 2007. https://doi.org/10.1007/s10338-007-0733-x

[7] J. Kaleta, Smart magnetic materials. Structure, manufacturing, testing properties, application (in Polish: Materiały magnetyczne SMART. Budowa, wytwarzanie, badanie właściwości, zastosowanie), Wrocław: Oficyna Wydawnicza Politechniki Wrocławskiej, 2013.

[8] J. Bomba, Damping in giant magnetostrictive material. Experiment, modelling, identification (in Polish: Tłumienie w materiale o gigantycznej magnetostrykcji. Eksperyment, modelowanie, identyfikacja), PhD thesis, supervisor: Jerzy Kaleta, 2010.

[9] J. Kaleta, K. Wachtarczyk and P. Wiewiórski, Multi-phase magnetostrictive actuator dedicated for applications using mechanical vibrations as controlled excitation, Experimental Mechanics of Solids - 28th Symposium, Jachranka, Poland, 2018.

[10] Codman Neuro, Post-operative Programming And X-ray Procedure Guide, 2015.

Keyword Index

About the Editors

Paweł Pyrzanowski, D.Sc., Ph.D., professor of Warsaw University of Technology, Faculty of Power and Aeronautical Engineering.
Member of many associations related to experimental mechanics:
- Polish Association of Experimental Mechanics – Founding member, Chairman of the Board since 2018;
- The Mechanics Committee of the Polish Academy of Sciences – Member, Chairman of the Experimental Mechanics Section – since 2016;
- Danubia-Adria Society for Experimental Mechanics – Member - representative of Poland since 2016;
- Symposium on Experimental Mechanics of Solids in memory of prof. Jacek Stupnicki – Chairman since 2012.

Author of more than 90 publications (monographs and chapters, scientific magazines and conference papers).

Mateusz Papis, M.Sc graduated from the Faculty of Power and Aeronautical Engineering of the Warsaw University of Technology in 2015.
Currently he works as an assistant at the Institute of Aeronautics and Applied Mechanics.
Member of Symposium on Experimental Mechanics of Solids Organising Committee since 2018.
His scientific interests include reliability and safety of systems, risk analysis.
Author of about 10 publication in these fields of study.

www.ingramcontent.com/pod-product-compliance
Lightning Source LLC
Chambersburg PA
CBHW071233210326
41597CB00016B/2033